SELECTED READINGS IN PHYSICS
General Editor: D. TER HAAR

ESSENTIALS OF LASERS

ESSENTIALS OF LASERS

by L. ALLEN

Lecturer in Experimental Physics, University of Sussex

THE QUEEN'S AWARD TO INDUSTRY 1966

PERGAMON PRESS

OXFORD · LONDON · EDINBURGH · NEW YORK
TORONTO · SYDNEY · PARIS · BRAUNSCHWEIG

PERGAMON PRESS LTD.,
Headington Hill Hall, Oxford
4 & 5 Fitzroy Square, London W.1
PERGAMON PRESS (SCOTLAND) LTD.,
2 & 3 Teviot Place, Edinburgh 1
PERGAMON PRESS INC.,
Maxwell House, Fairview Park, Elmsford, New York 10523
PERGAMON OF CANADA LTD.,
207 Queen's Quay West, Toronto 1
PERGAMON PRESS (AUST.) PTY. LTD.,
19a, Boundary Street, Rushcutters Bay, N.S.W. 2011, Australia
PERGAMON PRESS S.A.R.L.,
24 rue des Écoles, Paris 5e
VIEWEG & SOHN GMBH,
Burgplatz 1, Braunschweig

First edition 1969
Library of Congress Catalog Card No. 68–55560

Printed in Great Britain by W. and G. Baird Ltd., Belfast

08 013319 3 (flexicover)
08 013320 7 (hard cover)

To Barbara, Michael, Carol, Jenny and David

Contents

Preface

THE intention of this book is not to survey the contemporary state of knowledge concerning the laser—several recent volumes have done this adequately—but to outline the essential principles upon which laser action depends. The papers reprinted in Part II of the book have been chosen because, although they are not now the last word on a particular aspect of the subject, they each set into action a certain line of enquiry. The topics they discuss are the basic theory of lasers, resonator theory, gas lasers, crystalline solid lasers, semiconductor lasers and mode structure.

The treatment of the subject, then, is historical although the time scale is short. It was a temptation to employ a genuinely historical approach and to reprint Einstein's 1917 paper "On the Quantum Theory of Radiation" and Fabry and Perot's paper on their interferometer. However, the former has been published in another volume of this series (*The Old Quantum Theory* by ter Haar), and the latter is unrewarding for the amount of space it would occupy.

It is hoped that the book, and the papers it reprints and comments upon, will serve both as a general guide to the laser, and so be of interest to the average undergraduate, and also as an introduction to more serious study of this interesting field.

I wish to thank M. D. Sayers for reading and helpfully commenting on the manuscript; my wife for her help in checking and commenting on it in its various stages; Dr. D. ter Haar for his helpful comments and assistance, and Mrs. A. Millard who typed and retyped the manuscript so efficiently.

Finally, I would like to thank the authors of the papers reprinted in Part II of this book for so readily giving their consent to reproduce their original papers. My thanks go, also, to their publishers, the American Institute of Physics, Macmillan (Journals) Ltd., Akademie Verlag and the American Telephone and Telegraph Company for giving their permission.

PART I

CHAPTER 1

Introduction to the Laser

THE word laser is an acronym derived from *light amplification* by the *stimulated emission* of *radiation.* A laser is sometimes called an optical maser, but although this has the advantage of indicating that the laser and the maser are members of the same family, and that the laser works in the optical region of the electromagnetic spectrum, this name is not perhaps a good one. Taken literally the name means *optical microwave* amplification by the stimulated emission of radiation—a contradiction in terms. Consequently the word laser will be used throughout this volume.

Until the first laser was made by Maiman in 1961, the human eye has only observed spontaneous emission which occurs when an assembly of excited atoms lose their excitation energy by emitting at random moments in time essentially independent, randomly phased, wave packets of finite length. Constructive and destructive interference of these waves results in some mean amplitude being observed for the resultant emission.

Consider an idealized two-level atom capable of emitting frequency ν; then if the waves are randomly phased and of amplitude a_k, we may write for the complex amplitude of a particular wave,

$$\phi_k = a_k e^{i\delta_k}$$

where δ_k is the phase angle. Hence for an assembly of n atoms the resulting amplitude will be

$$\Phi = \phi_1 + \phi_2 + \dots + \phi_k + \dots + \phi_n$$

Now, $$\phi_1 = a_1 (\cos \delta_1 + i \sin \delta_1)$$

$$\phi_2 = a_2 (\cos \delta_2 + i \sin \delta_2)$$

and

$$\Phi = A\,(\cos \Delta + i \sin \Delta)$$

where $$A \cos \Delta = \sum_{1}^{n} a_k \cos \delta_k$$

and $$A \sin \Delta = \sum_{1}^{n} a_k \sin \delta_k$$

Square and add,

$$A^2 = \sum_{1}^{n} a_k^2(\sin^2 \delta_k + \cos^2 \delta_k) + \sum_{j \neq k} (a_k \cos \delta_k a_j \cos \delta_j + a_k \sin \delta_k a_j \sin \delta_j)$$

But the cross-terms will be zero on average since there is no correlation in the behaviour of the different waves. Then,

$$A^2 = \sum_{1}^{n} a_k^2 = n\bar{a}^2$$

if $\bar{a}$ is the mean amplitude of the disturbances. It should be noted that the resultant intensity is proportional to the number of excited atoms, n.

If in some way the atoms could be made to emit their waves in phase then the equivalent situation would be

$$\phi_1 = a_1\,(\cos \delta + i \sin \delta)$$

$$\phi_2 = a_2\,(\cos \delta + i \sin \delta)$$

and

$$\Phi = A\,(\cos \Delta + i \sin \Delta)$$

$$\therefore \quad A \cos \Delta = \cos \delta \sum_{1}^{n} a_k$$

and

$$A \sin \Delta = \sin \delta \sum_{1}^{n} a_k$$

$$\therefore \quad A^2 = n^2 \bar{a}^2$$

The resultant intensity is proportional to the square of the number of excited atoms, n^2.

The preceding discussion is idealized since no attempt is made to consider the finite spread of frequencies due to natural line-width, atoms perturbing each other and the movement of the atoms. It does, however, show that if the atoms could be made to emit in phase then the intensity of the emitted light would be much increased. The laser is, to some extent at least, the result of turning this possibility into a reality; although the effects of saturation ensure that the n^2 dependence is not, in fact, achieved.

For a laser to be possible it is essential that conditions be created whereby the atoms are stimulated to emit at frequency ν by the interaction of the atom with an electromagnetic field of the same frequency. When this occurs the quanta are no longer emitted randomly, but in a definite relationship to the stimulating field. This process is known as stimulated emission.

Although the acronym *laser* refers to light amplification the device is more an oscillator than an amplifier. Analogously with an electronic oscillator the constituent components of the laser may be thought of as a tuned circuit, a feedback system and a power supply. Consider an assembly of two-level atoms whose energy levels are radiatively connected by light of frequency ν. The relative population of these energy levels will be found to be a matter of some interest but, in general, if energy is supplied to the system some electrons will be excited to the upper of the two states, and light of frequency ν will be emitted as they return to the lower state. It may be shown that if these atoms are placed between two mirrors then under certain conditions of mirror alignment this emitted light can build up into a standing wave. Such a system is a resonant cavity and constitutes a tuned circuit resonating at frequency v.

Since the mirrors reflect the emitted light back into the atomic assembly they serve as a means of achieving feedback. Circuit losses are simulated by the fact that the mirrors will have less than 100% reflectivity and that diffraction losses will be introduced by the finite extent of the mirrors. Hence to maintain the strength of the electromagnetic field in the cavity, power must be put into the system to re-excite the atomic electrons to the upper energy level.

In this way radiation will continue at frequency v and the decaying amplitude of the cavity wall will grow again. It is this cavity wave which serves to stimulate emission from the atoms, provided certain other criteria are satisfied.

The nature of the resonator, the feedback mechanism and the frequency at which the oscillation takes place may be thought to be too far removed from the electronic equivalent for the analogy to be useful. However, provided the electronic oscillator chosen is of a suitable type, and in particular has certain non-linear characteristics, the equations describing the behaviour of the laser are very closely analogous indeed (see Van der Pol, 1934, and Lamb, 1964).

It will be seen in later chapters that the physical structure of the apparatus needed to create a suitable environment for stimulated emission to occur to a sufficient extent is such that the light is emitted into a small solid angle, unlike spontaneous emission which is emitted in all directions. Although a simple argument, this, together with the idea of the atoms emitting in phase, may be seen to be an explanation of the great intensity of the emitted laser beam.

In this introductory and greatly simplified view of the laser two of its characteristic properties are exposed. First its great intensity; and second the fact that unlike spontaneous emission which occurs in short unrelated wave-trains, and has a short coherence time, laser light has a high degree of temporal coherence. This means that for light from a particular part of the wavefront, it is possible to predict the phase and amplitude at any point as it travels over a large distance. This implies that the light is emitted in long wave-trains, as Fourier analysis shows, this in turn implies a high degree of monochromaticity. It is also known that the laser light has a high degree of spatial coherence. Spatial coherence means that there is a fixed phase relationship between the light at different positions across the beam. It should be made clear that this property does not follow from the previous simple discussion, and is not a result of stimulated emission. Both temporal and spatial coherence may be encompassed in the mutual coherence function (see Born and Wolf, 1964 chap. X) and to a great extent they are not strictly separable entities since, as a light beam propagates, wave-trains from different parts of the emitting aperture overlap and become inseparable. A great deal of caution must be exercised when discussing the coherence properties of the laser, and many problems remain to be resolved; at this level they are, perhaps, best avoided.

The wavelength range over which laser oscillation may be achieved extends from the quartz ultraviolet to sub-millimeter wave-lengths. The power in the resultant beams may be between a fraction of a milliwatt and tens of megawatts. Some lasers emit a continuous wave, others can emit only in pulses and a few can work in either manner. The atoms involved in the laser transition of frequency ν may be in a gas, a crystal, a glass, a semiconductor or a liquid. Each style of operation has its own characteristic attributes and each laser material its idiosyncrasies.

There are many applications of lasers and this might lead one to suppose that everything that needs to be known about the device is already known. This, however, is not the case and a great deal of fundamental research into laser processes remains to be done. What follows is very much by way of an introduction to the laser.

CHAPTER 2

The Theory of Lasers

CONSIDER a two-level atom, where the energy of the upper level is E_2 and that of the lower, E_1. If the levels are radiatively connected by frequency ν, and N_2 and N_1 are the number of atoms with their valence electrons in state E_2 and E_1 respectively, we may write, at thermal equilibrium:

$$\frac{dN_2}{dt} = -A_{21}N_2 - B_{21}N_2\rho(\nu) + B_{12}N_1\rho(\nu) = 0 \tag{1}$$

where $\rho(\nu)$ is the radiation density associated with temperature T, the temperature at which the system is thermalized and A_{21}, B_{12} and B_{21} are the Einstein coefficients.

From (1), we find for $B_{12} = B_{21} = B$

$$B\rho(\nu)(N_1 - N_2) = A_{21}N_2$$

$$\therefore \quad \frac{A_{21}}{B\rho(\nu)} = \frac{N_1 - N_2}{N_2} = \frac{N_1}{N_2} - 1 \tag{2}$$

Now the number of atoms radiating spontaneously per second will be

$$n_{\text{sp}} = A_{21}N_2 \tag{3}$$

and the number of stimulated emissions per second is

$$n_{\text{st}} = BN_2\rho(\nu) \tag{4}$$

Divide (3) by (4)

$$\frac{n_{\text{sp}}}{n_{\text{st}}} = \frac{A_{21}}{B\rho(\nu)} \tag{5}$$

From (5) and (2)

$$\frac{n_{sp}}{n_{st}} = \frac{N_1}{N_2} - 1 = \exp\left(\frac{h\nu}{kT}\right) - 1 \tag{6}$$

since at thermal equilibrium the Boltzmann distribution applies and

$$N_2 = N_1 \exp\left(-\frac{h\nu}{kT}\right)$$

In the optical region of the spectrum where $\nu \sim 5 \times 10^{14}$ Hz and for a moderate temperature $T \sim 450°$K, we find that

$$\exp\frac{h\nu}{kT} \sim e^5$$

Hence

$$n_{sp} \gg n_{st}$$

In other words, the probability of obtaining stimulated emission in the optical region of the spectrum is very small, since any excited state atoms are much more likely to radiate spontaneously. This discussion rests on the assumption that the atomic system is in thermal equilibrium, but a useful guide as to what is required to achieve stimulated rather than spontaneous emission may be obtained from equation (6).

It is true when

$$\exp\left(\frac{h\nu}{kT}\right) < \ln 2, \text{ that } n_{st} > n_{sp}$$

but, at optical frequencies, this corresponds to an exceedingly high value of temperature. In this case the number of stimulated photons will only just exceed the number emitted spontaneously. It will be shown later in this chapter that a small excess will invariably be insufficient to produce laser action. If, on the other hand, $N_2 > N_1$ then the system would not be in thermal equilibrium and n_{sp} is not necessarily greater than n_{st}. Consequently the first requirement for laser action must be that the energy levels concerned are not in thermal equilibrium, and that the upper of the two levels be more populated than the lower. This is known as population inversion.

Equation (4) shows that for a large amount of stimulated emission N_2 needs to be large, and so too does the energy density of the radiation field at the appropriate frequency ν. To achieve laser action, then, it is necessary to devise a means of achieving $N_2 > N_1$ and to build up the radiation field at frequency v. It is intuitively reasonable to suppose if the upper energy level has a fairly long spontaneous lifetime and the lower a short lifetime, that the chances of population inversion occurring are increased. A detailed discussion of how this non-equilibrium state may be achieved will be discussed in subsequent chapters.

Schawlow and Townes (1958) (Reprinted Paper No. 1) and Prokhorov (1958) (Reprinted Paper No. 2) independently discussed the conditions for achieving laser action. They each considered that although it was not possible to construct an optical resonator of the dimensions of the wavelength concerned, as may be done in the microwave region of the spectrum, the plane-parallel Fabry–Perot interferometer was, in fact, a suitable resonator. A Fabry–Perot interferometer is a more lossy type of cavity than a microwave cavity because it has no side walls, but both Schawlow and Townes, and Prokhorov thought that the main source of loss of energy would be due to the fact that the reflection coefficient of the mirrors could never be unity.

The light in the cavity may be considered to travel backwards and forwards between the mirrors, and the effective decay time of the cavity may be calculated by the simple expedient of considering that the energy of the wave falls to $1/e$ of its original value in n traverses, due to the imperfect reflectivity). Then if R is the reflection coefficient of the mirrors,

$$R^n = \frac{1}{e}$$

and, taking the logarithm and expanding,

$$n \approx \frac{1}{1 - R}$$

If the length of the cavity is d, the time, t_c, taken for the energy to decay to $1/e$ of its previous value will be

$$t_c = nd/c$$

$$\therefore \quad t_c = \frac{d}{c(1 - R)} \text{ seconds} \qquad (7)$$

If there are p modes effective in producing spontaneous emission where $p(\nu)\, d\nu$ is the number of modes between ν and $\nu + d\nu$ then, for an excess population $\bar{N} = N_2 - N_1$, to sustain a quantum in each mode we may write

$$\frac{\bar{N} h\nu}{p\tau} \geq \frac{h\nu}{t_c} \qquad (8)$$

where τ is the spontaneous lifetime of the upper energy level. That is, there are $\bar{N}$ quanta of energy $h\nu$ radiating into p modes in τ seconds. This rate per mode must be higher than the rate of decay of energy in the mode, $h\nu/t_c$. It is possible to write this since in a steady state the rate of stimulated emission in a single mode is just equal to the rate of spontaneous emission into the same single mode. Schawlow and Townes offered this statement without proof, but Yariv and Gordon (1963) showed that this is a special case of the fact that "the ratio of the induced transition rate to the rate of spontaneous emission into a given mode is equal to the number of quanta in the mode". Hence, from (8)

$$\bar{N} \geq \frac{p\tau}{t_c}$$

Now for a Doppler-broadened line Schawlow and Townes show that

$$p = \frac{8\pi^2 \nu^2 V \Delta\nu}{(\pi \ln 2)^{\frac{1}{2}} c^3}$$

where V is the cavity volume.

Hence,

$$\bar{N} \geq \frac{8\pi^2 \nu^2 V \Delta\nu \tau (1 - R)}{(\pi \ln 2)^{\frac{1}{2}} c^2\, d}$$

after substituting for p and t_c.

Hence the population inversion per unit volume is given by

$$\bar{n} \geq \frac{\bar{N}}{V} \geq \frac{8\pi^2\tau(1 - R)\nu^2\Delta\nu}{(\pi \ln 2)^{\frac{1}{2}} dc^2}$$

But the lifetime of the upper level (Condon and Shortley, 1935) is given by

$$\tau = 3hc^3/64\pi^4\nu^3\mu^2$$

where μ^2 is the dipole matrix element.

$$\therefore \quad \bar{n} \geq \frac{3}{8} \frac{hc\Delta\nu(1 - R)}{\nu\pi^2(\pi \ln 2)^{\frac{1}{2}}\mu^2 d}$$

and since for a Doppler broadened line

$$\Delta\nu = \frac{\nu}{c}\left(\frac{2kT}{m} \ln 2\right)^{\frac{1}{2}}$$

then

$$\bar{n} \geq \frac{3}{8} \frac{h}{\pi^{\frac{3}{2}}} \left(\frac{2kT}{m}\right)^{\frac{1}{2}} \frac{(1 - R)}{\mu^2 d}$$

Although this formula is derived on purely speculative grounds, it may be shown (Allen and Jones, 1967) that it yields good agreement for the value of the population inversion per unit volume obtained by Lamb (1964) in his rigorous semi-classical treatment of the problem. It should be realized for instance that the treatment presented here has tacitly taken over the Einstein coefficients for free space and applied them for the case of a cavity. A more sophisticated approach to the time constant of the cavity is possible if the "quality factor", or Q, of the cavity is invoked; the assumption of a wave being reflected back and forth is made unnecessary by expressing the problem in such terms. The above discussion has the advantage, however, of using the same language as Schawlow and Townes.

These relatively simple arguments serve to show that, given a medium with a sufficiently high population inversion, the electromagnetic field builds up so that the stimulated emission may be appreciably greater than the total spontaneous emission even in the case of a multi-mode cavity. As Schawlow and Townes realized, the large number of modes at infrared or optical frequencies which are present in any cavity of reasonable size poses problems. This is because of the need for a high level of spontaneous emission to ensure that sufficient photons go into any one particular mode, to maintain the rate of stimulated emission. Consequently they considered the possibility of selecting radiation from a single mode, and of making a multi-mode cavity sufficiently lossy to suppress the oscillations in unwanted modes. They considered how this might be done using two plane mirrors. Rather than consider their arguments in detail it is more fruitful to consider the work of Fox and Li (1961) (Reprinted Paper No. 3). Fox and Li established on the basis of diffraction theory that two mirrors of finite extent may be said to sustain certain modes of oscillation. The Fabry–Perot interferometer, as it has been known for the previous eighty years, turns out to be a particular case, and one where the apparently naive assumption of a plane wave being reflected backwards and forwards many times serves to yield a solution which approximates very well to what actually occurs.

It is worth noting, before these calculations are considered, that in terms of the simple semi-classical theory of Schawlow and Townes the need for a resonant cavity can be considered in one of two possible ways. It either constitutes a way in which a large radiation density may be made to grow at the appropriate frequency ν, or it may be thought, within the limits of the theory so far, as being a device for increasing the length of the active medium. In this way the small amplification coefficient may be exploited, and the intensity is able to increase at such a rate that it overcomes other inherent losses of light such as scattering or absorption. Such a theory will not, of course, allow for modes of oscillation nor for certain characteristic laser properties which are essentially the effect of a feedback mechanism. However, by writing the amplification coefficient in an analogous way to the more common absorption coefficient, considerable insight may be achieved concerning the desirable degree of population inversion. The amplification coefficient at the centre of a Doppler-broadened line, k_o, is given by (Mitchell and Zemansky, 1934)

$$k_0 = \left(\frac{\ln 2}{16c^2\pi^3}\right)^{\frac{1}{2}} A_{21} \frac{\lambda^4}{\Delta\lambda_D}\left(N_2 - \frac{g_2 N_1}{g_1}\right)$$

where A_{21} is the transition probability from level 2 to 1, λ the wavelength, $\Delta\lambda_D$ the Doppler width and g_1 and g_2 the statistical weights of the levels involved.

If $N_2 < (g_2/g_1)N_1$ then this is an absorption coefficient and if $N_2 > (g_2/g_1)N_1$ and there is population inversion, it is an amplification coefficient and may be associated with the gain/unit length of laser active medium.

It will be realized from this discussion that the laser is a noise-started oscillator where the noise is spontaneous emission. This randomly phased light with its finite spread of frequency serves to build up a radiation field at certain resonant frequencies, which, in turn, acts as the stimulating field in the induced emission process enjoyed by the other excited atoms.

The first of a whole series of important papers investigating the idea of resonant modes in an optical interferometer or cavity was that of Fox and Li. They did not consider the case of a cavity containing an active medium but only that of a passive cavity. As they pointed out, this formulation is satisfactory provided the interferometer is immersed in the active medium and there are no side-wall discontinuities. They considered a wave which propagated backwards and forwards between two parallel plane mirrors. An arbitrary initial field distribution was assumed at the first mirror, and the field at the second mirror was computed as the result of one transit. This distribution was then used to compute the distribution back at the original mirror as a result of the second transit, and so on. The distributions were computed on the assumption that the scalar theory of diffraction, as epitomized by the Kirchhoff intergal, could adequately describe the physical situation. This assumed, justifiably, that the wavelength of the light is small compared with the dimensions of the mirror, and that the field is transverse and polarized in one direction only.

The field at the mirror after one transit may be expressed as

$$u_p = \frac{ik}{4\pi} \int_S u_a \frac{e^{-ikR}}{R} (1 + \cos\theta)\, dS$$

where u_a is the field at the aperture, or in this case the first mirror, k the wave-number, R the distance from a point on the aperture to the point of observation and θ the angle of R to the normal to the aperture. After q transits the field at one mirror due to the reflected field from the other is given by replacing u_p by u_{q+1} and u_a by u_q. If a situation akin to a stationary mode of oscillation exists, then the distribution of the field at the mirrors would undergo negligible change from reflection to reflection, and would settle down after a certain number of transits to a steady state. If such a situation exists then the fields across the two mirrors become identical except for a complex constant; and

$$u_q = \left(\frac{1}{\gamma}\right)^q V$$

where V is the distribution function which does not vary from reflection to reflection and γ is the complex constant.

Substituting this condition in the relationship between u_{q+1} and u_q yields

$$u_{q+1} = \left(\frac{1}{\gamma}\right)^{q+1} V = \frac{ik}{4\pi} \int u_q \frac{e^{-ikR}}{R} (1 + \cos\theta)\, dS$$

$$= \frac{ik}{4\pi} \int \left(\frac{1}{\gamma}\right)^q V \frac{e^{-ikR}}{R} (1 + \cos\theta)\, dS$$

$$\therefore \quad V = \gamma \frac{ik}{4\pi} \int V \frac{e^{-ikR}}{R} (1 + \cos\theta)\, dS$$

$$= \gamma \int VK\, dS$$

Hence a function V which satisfies this equation may be regarded as a normal mode of the interferometer defined at the mirror surface, and $\ln \gamma$, which specifies the attenuation and phase shift the wave suffers during each transit, can be regarded as the propagation constant associated with this normal mode.

Fox and Li obtained numerical solutions for several geometric configurations including rectangular plane mirrors, circular plane mirrors and confocal spherical, or paraboloidal, mirrors. It is found that in each case there is a solution for V and that such an interferometer can be said to sustain various modes of oscillation. This implies that if one of these distributions is assumed to be introduced as an initial wave at one mirror, it will reproduce the same distribution at the other mirror. Such a distribution may be regarded as a normal mode of the interferometer. The lowest order mode, that is the mode of simplest symmetry, is found to have the lowest diffraction loss. It has a high intensity at the middle of the mirror, and low intensity at the edges. The diffraction loss due to over-spill around the edges of the mirrors is consequently much lower than would be predicted for a uniform plane wave. The same is essentially true, but to a lesser extent, for higher order modes, of more complex radial and angular symmetry. It may be noted in passing that, in spite of the success of the Fabry–Perot interferometer as an optical instrument, a plane wave does not resonate in the interferometer. It may be considered, however, to be decomposed into a set of modes which do resonate. In the usual case of the Fabry–Perot interferometer where the mirror separation d is small and the Fresnel number, $a^2/d\lambda$ where a is the mirror radius, is typically $\sim 5 \times 10^4$ the individual mode resonances are essentially degenerate and cannot be separately resolved. Hence a plane wave appears to resonate. In the case of a typical gas laser cavity the Fresnel number is $\sim 10^2$ and the degeneracy is removed.

The net result of this sort of analysis, as carried out by several workers in addition to Fox and Li (Boyd and Gordon, 1961; Boyd and Kogelnik, 1962; Collins, 1963), is that the resonant modes of a laser cavity can be categorized as TEM_{nmq} modes (see, for instance, Collin, 1960). Here n and m specify the angular and radial symmetries of the mode respectively, and q the fact that the round trip phase change of the wave in travelling from one mirror to the other and back again is $q.2\pi$.

In practice, the rigorous conditions applied by the degree of parallelism required between the two mirrors has meant that spherical mirrors are usually employed in a confocal or near confocal configuration. The diffraction losses of such configurations are low and the mirror alignment by no means as critical. A thorough survey of the theory of cavities is given by Kogelnik and Li (1966).

Now that it is established that it is possible to make a resonator at optical wavelengths, the theories of Schawlow and Townes and Prokhorov, whose paper may be seen to have followed essentially the same reasoning, may be reconsidered. To achieve oscillation it is clearly necessary for the gain per transit of the medium to exceed the losses per transit. The losses occur because of the absence of side walls to the cavity, diffraction losses around the sides of the mirrors and losses due to the less than perfect reflectivity of the mirrors. No mirror can ever be perfectly reflecting and indeed, so that some of the laser light may be released from the cavity to create a beam of laser light, it is necessary to have a finite transmission coefficient. The reflection losses are much more serious than the diffraction losses, and this has meant that for low gain laser systems very high reflectivities are necessary, $\sim$99·7%. To summarize: the reflection losses and medium gain determine whether or not the laser oscillates, and the diffraction losses determine the mode of oscillation.

The final service that Schawlow and Townes performed in their paper, which started so much research, was to consider the anticipated linewidth of a laser oscillator. Gordon, Zeiger and Townes (1955) had previously produced a formula for the bandwidth of radiation in the microwave region due to thermal radiation in the cavity. This was

$$\Delta\nu_M = \frac{4\pi kT}{P}(\Delta\nu)^2$$

where P is the power in the mode.

Analogously, Schawlow and Townes postulated that spontaneous emission effects were equivalent to a thermal noise temperature T where

$$T = \frac{h\nu}{k}$$

and so,

$$\Delta\nu_L = \frac{4\pi h\nu}{P}(\Delta\nu)^2$$

where $\Delta\nu$ is the bandwidth of the passive cavity resonance. This suggests ideal theoretical linewidths for lasers in the range 10^{-1} to 10^{-7} Hz. The problems of stabilizing the actual oscillation frequency are such as to render such values of linewidth relatively meaningless. The theoretical formula quoted above is, however, within a factor of 2 of those derived by more complex approaches (Lamb, 1965).

CHAPTER 3

Crystalline Solid Lasers

MAIMAN'S deceptively brief paper in Nature in 1960 (Reprinted Paper No. 5) heralded the arrival of the laser as an entity. In about 300 words he analysed how population inversion had been achieved in ruby, the wavelength of the laser emission, how the resonant cavity was made, and the fact that the emitted radiation had a much narrower bandwidth than its spontaneous equivalent.

The required optical cavity was made by cutting the ends of the ruby rod accurately plane and parallel and silvering them. Subsequently more sophisticated techniques have been employed such as depositing multi-layer dielectric coatings, or coating only one end and cutting a wedge on the far end of the rod to act as a roof prism end reflector. Alternatively a pair of external mirrors may be used, allowing other optical components to be placed in the cavity.

Maiman found that if the ruby is irradiated with light at a wavelength of about 5500 Å, the triply ionized chromium atoms which dope the aluminium oxide and give ruby its characteristic red colour, become excited to the $^4F_2$ state (see Fig. 3.1). These excited ions then decay quickly, via phonon-assisted non-radiative transitions, to the pair of levels associated with the ^2E state. These in turn decay radiatively to the ground state as the R_1 and R_2 resonance fluorescence lines of ruby. This decay rate is, however, slow, and in the presence of intense excitation at 5500 Å the population of ^2E becomes greater than that of the ground state. The required inversion of population is then achieved and noise-started laser oscillation may be observed at 6943 Å.

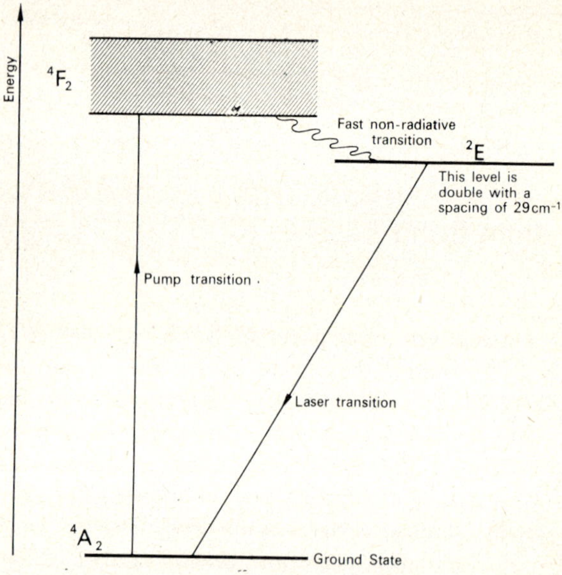

FIG. 3.1. Energy-level diagram of Cr^{3+} ions in Al_2O_3, ruby.

The pump light for such a system must come from a flash tube, since a very intense pulse of light is needed to achieve the desired level of pumping. A large proportion of the light is taken up because of the broad absorption band of ruby. The way in which this light is coupled into the ruby rod has been the source of a great deal of study and several techniques are available. The flash tube may be in the form of a helix and the ruby threaded through it, or a number of small flash tubes may be packed closely around the ruby; but a more elegant form of coupling, which allows the ruby to be more easily temperature controlled, is that of the elliptical cylinder. In this method a cylinder of elliptical cross-section is made with a highly reflecting inner wall. The ruby rod with its axis parallel to that of the cylinder is placed such that it occupies one of the foci of the elliptical cross-section. A long straight flash tube is then placed in the position of the other focus. All the light from the flash tube is then focussed on to the ruby rod, but air can circulate around the rod and to some extent control its temperature. The effect of temperature changes are twofold, first they tend to shift the frequency of laser oscillation since the length of the cavity changes, and second if the temperature becomes too high the distribution of Cr^{+++} ions in the appropriate energy levels is such that the creation of a population inversion is inhibited.

Maiman (1960a) (Reprinted Paper No. 4) had already investigated the optical properties of ruby in an earlier paper. He had found that the effective lifetime of the ^2E level was about 5 msec, and that the lifetime was almost exclusively due to spontaneous emission rather than any other decay process.

The decay rate between $^4F_2$ and ^2E he found to be about 2×10^7/sec, and the number of fluorescent quanta emitted in the transition from ^2E to the ground state $^4A_2$, very nearly equalled the number absorbed at about 5500 Å by the $^4F_2$–$^4A_2$ transition. Maiman was able to show that population changes in the ground state of ruby due to optical excitation could be observed. He mounted the crystal between parallel silvered plates to form a microwave cavity, resonant at the ground-state zero-field splitting of 11·3 GHz. The fact that the Q increased when a small magnet was brought near the ruby so that the levels were no longer separated by 11·3 GHz, showed that absorption by magnetic dipole transitions accounted for about half the cavity losses. A 200 μsec pulse of light irradiated the crystal, and the microwave magnetic absorption was observed to decrease rapidly and then return to the equilibrium value with a time constant of about 5 msec. Maiman attributed this to temporary depletion of the ground state population with subsequent decay from the fluorescent level, ^2E. Finally he showed that absorption by the $^4F_2$–$^4A_2$ band decreased for about 5 msec after the crystal has been illuminated with light from the flash-tube, thus confirming his explanation.

It may be seen from this that Maiman had carefully analysed ruby to determine whether or not the required criteria for laser oscillation could be satisfied and, having ascertained that there was a good chance of success, then made the ruby laser. This laser, although exceedingly useful and powerful, is in some respects a crude device. The lower level of the laser oscillation is the ground state and this will always have a large population at thermal equilibrium. Consequently rather a lot of energy has to be put into the system before the ^2E state can be inversely populated with respect to the ground state. In practice 10^{18}–10^{19} ions are needed in the metastable state before inversion can be achieved. It so happens that producing energy at 5500 Å from a flash tube is not especially hard, but a good deal of energy is released at other frequencies as well and much electrical energy is wasted in its creation. The spurious light must also be dissipated somewhere and tends to warm up the crystal.

It would appear that ruby must be inherently a laser which can work only in pulses, since a large input pulse of light is necessary to create momentarily an inversion between ^2E and $^4A_2$. However, Nelson and Boyle (1962) have demonstrated that it may be made to work continuously. This was achieved by very careful design of the way in which the pump light was introduced to the crystal, and by placing the crystal in a dewar of liquid nitrogen to remove the unwanted heat. The output was found to consist of a series of spikes lasting a microsecond or so, each separated by a few microseconds. Since this type of output may be continued indefinitely, it is said to be continuous. This is in contrast to Maiman's pulsed laser, which emitted similar "spikes", but only for $\sim 0{\cdot}5$ msec at a time. This does not detract from the fact that, because the ground state is involved, the ruby laser will always be inefficient.

Sorokin and Stevenson (1960) (Reprinted Paper No. 6) described stimulated emission in a different atom and host crystal. This was from trivalent uranium ions substituted for divalent calcium ions in calcium fluoride. The oscillation takes place on a transition from a metastable state to a level approximately 515 cm^{-1} above the ground state. At low temperatures this state is less populated than the ground state by at least a factor of 10^{10}, and hence the population of the upper laser level does not need to be as large as it was in ruby, and it may be much less than the number of ions in the ground state. Because of this the required pump power is a factor of 500 down compared with ruby.

The doped calcium fluoride rod was prepared in the same way as ruby to obtain an optical cavity and similar techniques were employed to inject the pump power. The experimental arrangement was rather more complicated since the sample crystal was inside a liquid helium dewar and the flash lamp excited it through a side port in the dewar. It may be noted that Sorokin and Stevenson observed that the linewidth of the laser radiation was narrow, "substantially less than the resolving power of the monochrometer", and that the angular spread of the beam was small.

There are now many crystalline solid laser systems embracing a large range of host lattices and active ions. The two systems discussed here are not simply the oldest solid-state lasers but also serve as two important prototypes which are often categorized as "three-level" and "four-level" laser systems. In the three-level system the reservoir of atoms to be pumped is also the lower laser level; in the four-level system neither of the levels involved in the pumping process is a laser level. Rate equations describing the two types of crystal laser have been investigated by Maiman (1961) (Reprinted Paper No. 7) and serve as a basis for understanding the excitation processes involved. Since the ground state of ruby is split it might appear that it could operate as a four-level system, but the splitting is too small for the upper of the ground-state levels to be appreciably depopulated.

By writing the rate equations connecting the energy levels involved in either the three-level or four-level systems, Maiman shows that it is possible to obtain expressions for the required pump power and effective temperature of the exciting source, in terms of the decay rates, ion density and involved frequencies of the particular system under discussion.

In the energy level arrangement shown in Fig. 3.2, the ground state is the terminal level for the transition $2 \rightarrow 1$. At least half of the ground state atoms must be excited into level 2 if population

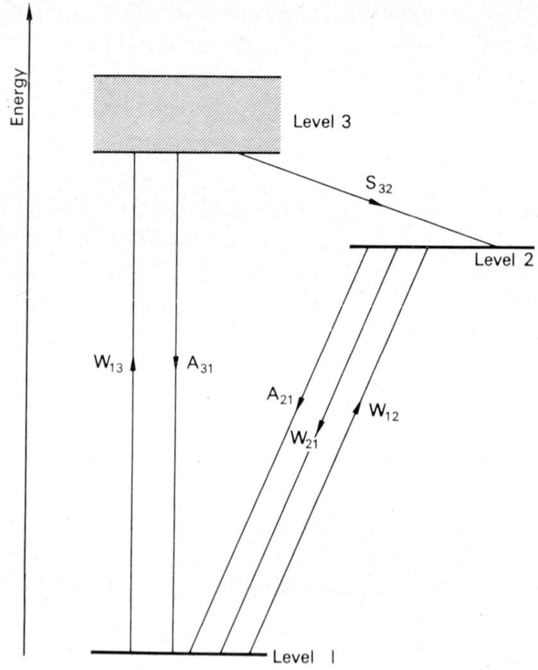

FIG. 3.2. Energy-level diagram for a three-level fluorescent solid.

inversion is to be achieved. This, as already mentioned, is the prevailing characteristic of the three-level system, Maiman writes the rate equations connecting the populations of the energy levels concerned as

$$\frac{dN_3}{dt} = W_{13}N_1 - (W_{31} + A_{31} + S_{32})N_3 = 0$$

$$\frac{dN_2}{dt} = W_{12}N_1 - (A_{21} + W_{21})N_2 + S_{32}N_3 = 0$$

$$N_1 + N_2 + N_3 = N_0$$

at the steady state.

W_{13} is the stimulated transition probability per unit time caused by radiation of frequency ν_{13}, W_{21} the stimulated transition probability due to radiation at frequency ν_{21}, A_{31} and A_{21} are the Einstein A coefficients and account for spontaneous emission, and S_{32} is the transition probability for the non-radiative transition $3 \rightarrow 2$. N_i refers to the population of level i. These equations may be solved and if, as in ruby, $A_{31} \ll S_{32}$ and $W_{31} \ll S_{32}$ then $N_3 \ll N_1$ and $N_3 \ll N_2$. The solution of the above equation then simplifies to

$$\frac{N_2 - N_1}{N_0} \approx \frac{W_{13} - A_{21}}{W_{13} + A_{21} + 2W_{12}}$$

To obtain stimulated emission at frequency ν_{21}, $(N_2 - N_1)$ must be positive. Hence it may be seen that $W_{13} > A_{21}$ is a necessary condition for laser oscillation. It is not a sufficient condition because $(N_2 - N_1)$ must be large enough to overcome the losses of the system. Maiman considers a wave travelling back and forth along the crystal, reflected by a mirror of reflectivity r at each end. Following similar ideas to those of Schawlow and Townes he deduces that at threshold,

$$\frac{N_2}{N_1} \approx 1 + \frac{2(1 - r)}{\alpha_0 l}$$

where l is the length of crystal, and α_0 is the normally measured absorption coefficient for the transition $1 \rightarrow 2$ under low power excitation. If the pumping light has power P, then

$$P_0 = nh\nu_p$$

where n is the number of incident photons per second in the region of the absorption band and ν is the mean frequency.

Also,

$$W_{13} = \frac{n\sigma_p}{A}$$

where σ_p is the absorption cross-section for the pump transition $1 \rightarrow 3$ and A the area of the crystal perpendicular to the incident beam. Consequently when $W_{13} = A_{21}$ and inversion just commences, the required light flux density, F, is given by

$$F = \frac{P_0}{A} = \frac{W_{13}h\nu_p A}{\sigma_p A} = \frac{W_{13}h\nu_p}{\sigma_p} = \frac{A_{21}h\nu_p}{\sigma_p}$$

The energy density in the crystal is then computed from this value of flux, although assumptions have to be made here to allow for the coupling between a pump source and an active crystal. In general, high pressure lamps have a more or less continuous spectrum, and it is not unrealistic to consider that the active medium is immersed in isotropic black-body radiation.

Assuming that the crystal is optically thin, that is the radiation density is uniform throughout its volume, Maiman writes,

$$W_{13} = \int \frac{c\sigma(\nu)\rho(\nu)\,d\nu}{h\nu_p \varepsilon^{\frac{1}{2}}}$$

where $\rho(\nu)$ is the energy density per unit volume per unit frequency interval and $\sigma(\nu)$ the frequency-dependent absorption cross-section. Since

$$\rho(\nu) = \frac{8\pi h\nu^3 \varepsilon^{\frac{3}{2}}}{c^3(e^{h\nu/kT} - 1)}$$

he is able to show that for $N_2 = N_1$ and $W_{13} = A_{21}$

$$T_p = \frac{h\nu_p}{k \ln\left[1 + (A_{31}/A_{21})\right]}$$

This gives the effective pump temperature, and hence is a measure of the power per unit area per unit frequency interval required. Comparison of the pump powers needed show that, as expected, the requirement in the four-level case is much less stringent. The paper concludes with a discussion of the fact that the presence of the crystalline material will introduce "light pipe" modes of propagation due to internal reflection and that strains, inhomogeneities and deviations from single crystallinity will all introduce distortion into the idealized Fox and Li modes. Consequently values of the required inversion calculated using the foregoing theory may not be sufficiently accurate when applied to a real crystal. Small corrections, therefore, need to be made to the values of pump power and effective temperature obtained.

The theory outlined here, is no more sophisticated than that of Schawlow and Townes; essentially an equilibrium value is assumed for the radiation density. A really detailed study of the effect of stimulated emission on such a system in the presence of discrete modes of oscillation will, and indeed does, predict other important results. Maiman's type of analysis is very useful because, in a simple way, a very good guide can be obtained to the sort of values required for ion density, pump power and so on.

The crystal laser has certain distinct advantages over other types of laser. In general, the longest lived excited metastable energy levels exist in impurity doped solids, and so a high degree of energy storage is possible. The density of active impurity ions is $10^{17}-10^{20}$ ions/cm^3 compared with 10^{15}–10^{17} atoms/cm^3 in gases, and so fairly high C.W. powers are possible. For a given output power a smaller volume of active medium is necessary than that required by a gas laser. To be a successful laser material a crystalline solid should have a narrow fluorescent linewidth, few impurities and scattering centres, and a broad strong absorption band coincident with the output of available types of flash tube. The upper laser level should have an essentially radiative lifetime and this should preferably be long. Smith and Sorokin (1966) review crystalline solid lasers especially well in their book.

CHAPTER 4

Gas Lasers

WHEN an atom, or ion, is in the environment of the host lattice of a crystalline solid, its energy level structure is fairly simple. The effect of the crystal field is to make a large number of free ion energy levels essentially degenerate and lose their identity. The essential simplicity of the level structure makes analysis of the Maiman type straightforward in principle, although it should be realized that lifetimes and cross-sections are often not well enough known for useful calculations to be made concerning a real system.

The energy levels associated with an atom or ion in the gaseous state are more numerous, and analysis of population processes is, correspondingly, more complex. Population of energy levels may be due to direct electron impact, transfer of excitation energy by a collision of the second kind, radiative cascade processes, dissociation, selective recombination or optical pumping. One or many of these processes may account for the relative population of one level to that of another. Certain rules follow from the solid-state case. If population inversion is to be achieved, not only must a device be found for selectively populating the upper of the two levels, but it will help if the effective lifetime of the upper level is long compared with that of the lower.

Schawlow and Townes (1958) investigated the possibility of selectively populating the $5p\ ^2P_{\frac{1}{2},\frac{3}{2}}$ levels of atomic potassium by optical pumping using radiation at 4047 Å. This wavelength corresponds to a resonance line from $5p\ ^2P$ to the ground state of potassium and could be obtained from a potassium lamp. If this light is directed into potassium vapour, some ground state atoms would be excited by absorption to $5p\ ^2P$, while the other levels would remain essentially unpopulated. The discussion is well worth reading, and the idea a good one. A similar idea was succesfully exploited using cesium vapour (Jacobs, Gould and Rabinowitz, 1961), but potassium has never been made to lase and further consideration of this system is not fruitful.

One of the first papers concerning the creation of population inversion in gases was that of Javan (Reprinted Paper No. 8). It should be explained that his phrase "negative temperature", which has rather died out in laser literature, merely means that if the Boltzmann distribution were taken to apply to a pair of levels where population inversion existed, the characteristic absolute temperature would have to be negative, since

$$N_{\text{up}} = N_{\text{low}} \exp\left(\frac{-h\nu}{kT}\right).$$

He assumes that in a gas discharge the main source of population of an excited state is due to collisions of the first kind, or direct electron impact, between energetic electrons and ground state atoms. In this case a free electron collides with a ground state atom, gives up some of its kinetic energy, thereby exciting the atom, and goes on its way with diminished velocity. An investigation of this process, ignoring for the moment the fact that electrons will also collide with excited atoms and change their excitation energy, yields interesting order-of-magnitude results.

Javan considers a three-level atom, the levels of which may be labelled 0, 1 and 2 where '0' refers to the ground state. Let $1/\theta_{0i}$ be the rate of excitation of the ith excited state by electron impact with the atoms in the ground state, and $1/\theta_{i0}$ the rate of the reverse process. If the electrons have a Maxwellian distribution then,

$$\frac{\theta_{0i}}{\theta_{i0}} = \exp\left(\frac{E_i}{kT_e}\right)$$

Provided $\theta_{i0} \ll \tau_i$, the radiative lifetime of the ith excited state, the level population will be due entirely to electron collisions and

$$\frac{N_i}{N_0} = \exp\left(\frac{-E_i}{kT_e}\right)$$

which means that there is a Boltzmann distribution of population corresponding to an electron temperature T_e. For $\theta_{i0} \gg \tau_i$, however, one obtains

$$\frac{N_i}{N_0} = \frac{\tau_i}{\theta_{i0}} \exp\left(\frac{-E_i}{kT_e}\right)$$

Since $$N_1 = N_0 \frac{\tau_1}{\theta_{10}} \exp\left(\frac{-E_1}{kT_e}\right)$$

and $$N_2 = N_0 \frac{\tau_2}{\theta_{20}} \exp\left(\frac{-E_2}{kT_e}\right)$$

then $$\frac{N_2}{N_1} = \frac{\tau_2}{\tau_1}\frac{\theta_{10}}{\theta_{20}} \exp\frac{-(E_2 - E_1)}{kT_e}$$

It is clear, under suitable conditions of τ_i and θ_{i0}, that N_2 may be made greater than N_1, although, as Javan observes, excessively large electron densities must be avoided or else $\theta_{i0} \gg \tau_i$ will be violated. Gas lasers depending almost exclusively on direct electron impact have since been made.

In the same paper Javan considers the transfer of excitation energy between excited states of two different atoms in a gas mixture. If there exists an energy level in one type of atom, *a*, of very nearly the same energy as a level in the other type of atom, *b*, then a large cross-section exists, in general, for an inelastic collision resulting in the transfer of excitation energy from atom *a* to atom b. If the appropriate level in gas *a* is metastable, that is it is for instance, unable to radiate downwards to any other levels without violating selection rules, or it is radiation-trapped (Holstein, 1947), it is likely to be quite heavily populated. If such an atom collides with a ground-state atom of type *b* its excitation energy may be transferred to it. Thus atom *b* is excited to a level which is probably not metastable, and atom *a* returns to its ground-state. In this way atom *b* may acquire a considerable population in that energy level, and, hopefully, be so populated that inversion exists between it and another level to which it is radiatively connected. Javan discusses such possibilities and in 1961 with Bennett and Herriott successfully used the process (Reprinted Paper No. 9) to create the first gas laser.

The first gas laser transition was achieved in a mixture of helium and neon. There is, unfortunately, a confusion of terminology in Javan's paper which should be explained. The neon energy levels are labelled with a now defunct nomenclature due to Paschen. The levels are best described in terms of Racah's $j - l$ coupling scheme, and many of the levels may be adequately described in terms of L–S coupling. The three terminologies may be compared by consulting Moore's *Atomic Energy Levels*, vol. I (1949), or by the energy level diagrams shown by Allen and Jones (1967). The remainder of this discussion will use Paschen notation to stay consistent with Reprinted Paper No. 9.

When a helium atom with its valence electron in the metastable state 2^3S_1 collides with a neon atom in its ground state, the excitation energy may be transferred to the $2s$ levels of neon. As may be seen in Fig. 4.1, the energy gap between 2^3S_1 and the $2s$ levels is small and so the cross-section for such a transfer of energy is large. It is known for energy gaps ("energy defects") which are greater than $\sim 800 \text{ cm}^{-1}$ that the cross-section for such processes is zero, and consequently the $2p$ levels will not become populated in this way. This provides, then, a selective population mechanism which prefers the $2s$ levels to those of $2p$. The question of whether or not the rate of population is sufficient to maintain a large enough population inversion to achieve laser action can only be answered in terms of the lifetimes of the levels involved. Javan *et al.* (1961) show that the lifetimes of the $2s$ levels are about one order of magnitude longer than those of the $2p$ levels. This is so even though $2s_2$ and $2s_4$ are radiatively connected to the ground state, and it would be reasonable to suppose that they were therefore short-lived. These transitions, however, undergo complete resonance

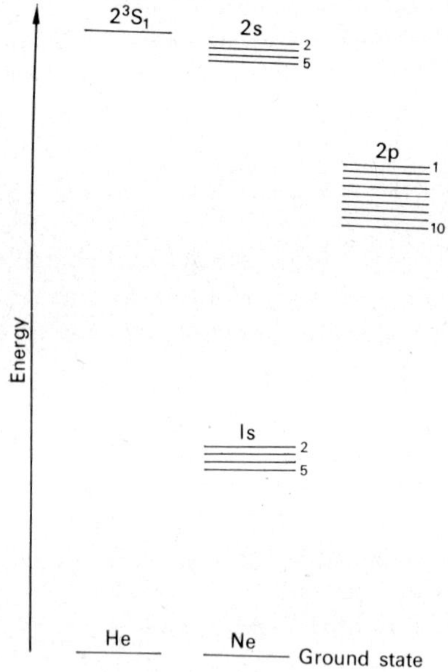

FIG. 4.1. Energy-level diagram of neon showing the coincident helium metastable level 2^3S_1.

trapping. This means that every time a photon is emitted, another ground state atom absorbs the photon and re-excites the state. This process can continue time and time again because of the large ground state population. The lifetimes of these levels are in the end determined not by their resonance transitions but by their decay to the $2p$ levels.

The total, and partial, pressures of the gas mixture are not arbitrary. The correct conditions have to be found to ensure sufficient transfer of excitation energy to the $2s$ levels, and the resonance trapping of the $2s$ to ground state transitions. A means has to be devised, too, of keeping the population of the $1s$ levels at a minimum. This last requirement is to ensure that the $2p$ levels are not excited by inelastic collisions between electrons and the metastable $1s$ levels. If this occurs the inversion between $2p$ and $2s$ will be destroyed. There is a very real danger of this process occurring, because the $1s$ levels are either radiation trapped or else have no allowed downward transitions. Consequently they tend to build up a sizeable population. This has meant that at least one dimension of the laser discharge tube must be small, so that these metastable atoms may collide with the wall of the tube and so depopulate.

There can be no doubt of the process creating the population inversion. When the He–Ne discharge is switched off, the $2s$ levels are found to decay with a time variation identical to that of the helium 2^3S_1 metastable state. Also, a larger value of amplification is found in the medium during the afterglow of the discharge than during a current pulse. This is because during the afterglow there are no electrons to excite $1s$ atoms to $2p$ by inelastic collisions.

Javan found that several different wavelengths could be made to oscillate corresponding to different $2s$–$2p$ transitions. The strongest transition was at 1·153 μ. Javan's laser consisted of an 80 cm long quartz tube, of internal diameter 1·5 cm and containing 0·1 torr Ne and 1 torr He. The optical cavity consisted of two plane-parallel reflecting mirrors placed within the gas tube via flexible bellows, so that they might be externally tuned. The discharge was excited by means of external electrodes connected to a 28 MHz R.F. generator. The output light has a power of a few milliwatts and the input power was around 50 W.

It might be noted that advances were very quickly made on this type of experimental technique. The difficulty of aligning a Fabry–Perot interferometer with an etalon spacing of 80 cm is considerable. As already stated in Chapter 2, the alignment is much less critical if a pair of spherical mirrors are used. Manufacture of the device is considerably simplified if the mirrors are external to the discharge tube. This was soon shown to be possible by the use of Brewster angle windows on the discharge tube. For a window of refractive index n, light incident at the Brewster angle, given by

$$\tan \theta_B = n$$

has that component which is polarized in the plane of incidence, totally transmitted. This means that for light polarized in this plane, multiple transits through such a window will not constitute a source of loss. This is unlike the case of a window perpendicular to the axis of the discharge tube, which would reflect ~4% of the light incident upon it at each pass. Here, unless the gain of the active medium were exceedingly high, the system would have more loss than gain, and would not be able to support oscillation. This has meant that, except for special systems where it is not wished to superimpose a polarization artificially, most gas lasers have sealed-off discharge tubes terminated at each end with a window at the Brewster angle.

The divergence of the ouput beam of the 1·153 μ laser was measured by intercepting the beam by the photocathode of an image convertor and photographing the then visible image. A screen containing nine slits was placed across the full ~1 cm aperture of the beam and a diffraction pattern resulted with very small minima. This indicates a high degree of spatial coherence. If the beam were totally incoherent no pattern would have resulted; and zero intensity at the minima corresponds exclusively to the case of perfectly spatially coherent light (see Born and Wolf, 1964, chap. X).

The remainder of the paper by Javan *et al.* is concerned with the measurement of the linewidth of the emitted radiation. It was realized that the linewidth would be many orders of magnitude narrower than could be measured with the resolution available using standard optical techniques. The photo-multiplier is a perfect square law detector and responds to beat frequencies between adjacent optical frequencies. Hence an attempt was made to observe the spectral distribution of the light by beating all the components of the line together. Using a frequency analyser, a strong signal was found at zero frequency, corresponding to all components beating with themselves, with a half-width of about 100 kHz. Further examination showed the presence of other beats centred on non-zero frequencies. It is perhaps worth while at this point to consider the work of Herriott (1962) who describes similar measurements on the same laser.

Herriott reviews, rather more critically than Javan the optical aspects of the device. In particular he is concerned with the interferometer structure, and the problems of reflector alignment. The energy distribution, and the degree of spatial coherence across the beam is examined. Finally, and this is where the paper is especially interesting, the spectral linewidth, cavity modes and mode-coupling are discussed. From the theory of optical resonators discussed in Chapter 2, it is known that axial modes will occur when

$$2d = q\lambda$$

where q is an integer

and so $$\frac{1}{\lambda} = \frac{q}{2d}$$

But $c = \nu\lambda$ $$\therefore \quad \nu = q\frac{c}{2d}$$

therefore adjacent modes will be separated by

$$\Delta\nu = \frac{c}{2d}$$

For $d \sim 1$ metre, then $\Delta\nu \sim 150$ MHz. The gain curve of the $1{\cdot}153\ \mu$ laser line is the Doppler-broadened profile of the transition in neon gas. This is from 1 to 2 GHz wide and consequently it is clear that a large number of axial modes will attempt to resonate. In the presence of loss in the cavity some of the gain will not be available to contribute to oscillation. It may be seen from the loss line, corresponding to losses f, in Fig. 4.2 that this means some axial modes will be below threshold. Nonetheless, a large number of modes will be able to oscillate. At any resonance, the laser linewidth that occurs is the result of the interaction between the combined optical transition and cavity linewidths (see Fig. 4.2).

Upon oscillation, the ouput line is found to be narrower than the cavity bandwidth, but it draws energy from a natural linewidth in the gas; atoms within a natural linewidth of the cavity resonance

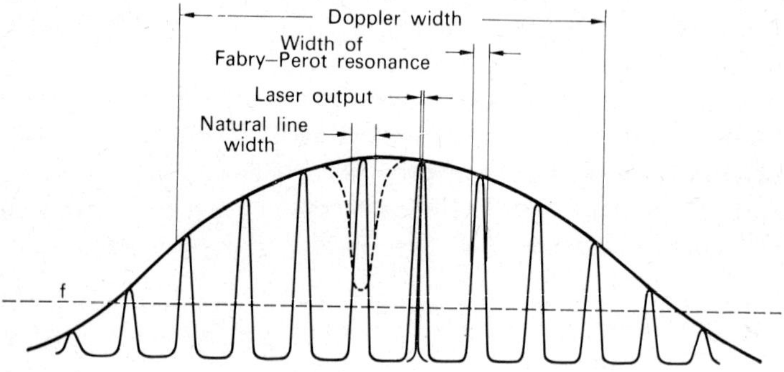

FIG. 4.2. A comparison of some important linewidths.

will be stimulated to emit. Since, argues Herriott, the spacing of the modes is 150 MHz and the linewidth is an order of magnitude smaller, it is possible that two adjacent modes will not need the same atoms to sustain them, and that both will oscillate. Examination of the ouput signal from a photomultiplier placed in the laser beam, using a spectrum analyser, showed two major peaks at 148 MHz and 296 MHz. This indicated that there were at least three separate longitudinal, or axial, modes oscillating at once. At 1.3 MHz to each side of these peaks there was a second peak. The calculations of Fox and Li show, for the cavity used, that the first off-axis, or unsymmetric mode will occur at 1·3 MHz from the axial one. These measurements demonstrated extremely well the existence of "Fox and Li" -type modes. The bandwidth associated with the beats yields information concerning the output linewidth of the radiation. The natural bandwidth of the spectrum analyser dictates the lower limit of such a measurement, but at that time it was clear that the linewidth was no more than a few hundred Hertz. More recent measurements show a few parts of a Hertz.

This behaviour is characteristic of gas lasers because the Doppler-broadened transition line is said to be inhomogeneous. This means that when oscillation occurs at one frequency the stimulated emission burns away all the available gain at that frequency down to the loss level (see Fig. 4.2) but does not affect the gain at other frequencies. This is because the probability of collision between excited atoms of different velocities is much less than the probability of their radiatively emitting. Consequently there is no transfer of energy from one part of the gain curve to another in such a transition. This is usually not the case in a solid state laser, where depletion of gain at one frequency often has the effect of an overall depletion in the gain curve. These effects are discussed again in Chapter 6.

Many aspects of the He–Ne laser are characteristic of any gas laser. Narrow linewidth, many modes available, the need for careful choice of partial pressure and total pressure of the gases, high reflectivity mirrors, high degree of spatial coherence, are some of the properties and experimental necessities. There are now a large number of different types of gas laser and not all their properties are shared by He–Ne. If there is no problem of metastable states that need to be depopulated, the requirement that one tube dimension be small is not a necessary one. Very few gas lasers use a R.F. generator as a source of input power. The power may be fed in as D.C., or in the form of so-called D.C. pulses which occur when a high power pulse is fed, via internal electrodes of fixed polarity, into the discharge region. The gain in some gases at some frequencies, is so big that powers as high as watts, or tens and even hundreds of watts may be emitted. The restraints upon the accuracy of setting of Brewster angle windows, of mirror reflectivity and alignment then becomes much less rigorous. The gas laser field has been extensively reviewed by Allen and Jones (1967) and by Garrett (1967).

CHAPTER 5

Semiconductor Lasers

AIGRAIN (1958) suggested that an infrared laser might be made using the radiative recombination of electrons and holes across the gap of a semiconductor. The same rules apply here as in the case of crystalline solid or gas lasers. The population of two electronic levels must be such that the upper level may be induced to emit a photon of energy $h\nu$, equal to the energy separation between the states, more often than the lower level absorbs photons. A resonant structure must be possible so that sufficient stimulated emission may occur for oscillation to commence. Some features of semiconductor lasers, though, are quite different from those of other lasers; these include a continuous spectrum of electronic states, and the type of pumping mechanisms used.

Semiconductors without impurities (intrinsic semiconductors) are insulators at low temperatures. At room temperature a number of binding electrons have enough thermal energy to be raised from the normally highest filled band, or valence band, across the narrow forbidden gap into the lowest normally empty band, the conduction band. These conduction band electrons can carry a current through the solid. The empty states in the valence band, or holes, can also conduct. These holes, in the presence of an electric field, act as positive charge carriers and move in the opposite direction to the electrons.

Bernard and Duraffourg (1961) (Reprinted Paper No. 10) investigated the possibility of obtaining stimulated emission in semiconductors from transmissions between the conduction and valance bands, and between one band and an impurity level. Consideration of what is meant by an impurity level will be deferred until later in the chapter. Bernard and Duraffourg point out that, in the one-particle approximation, an electronic state in a solid may be represented by a Bloch wave defined over the whole crystal. Two states are assumed. The first, corresponding to the valence band, with a wave vector $\boldsymbol{k}_i$ and energy $E_v(k_i)$ and the second, corresponding to the conduction band, of wave vector $\boldsymbol{k}_j$ and energy $E_c(k_j)$. Figure 5.1a indicates the density of states against energy for a pure semiconductor.

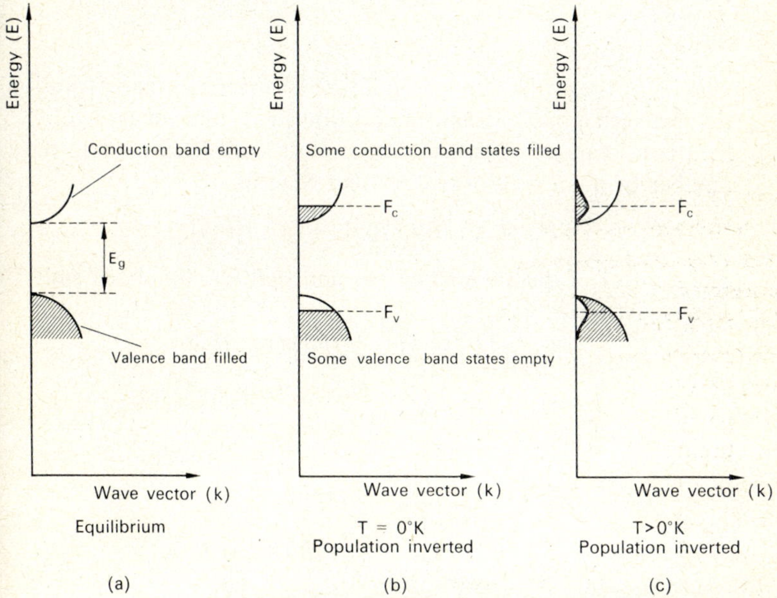

FIG. 5.1. Distribution of population for the conduction and valence bands in a semiconductor.

Under equilibrium conditions the occupation probability of any state of energy $E(k)$ is given by the Fermi–Dirac function,

$$f = \frac{1}{1 + \exp\,[E(k) - F_0]/kT}$$

where F_0 is the Fermi level for the electrons, that is, the energy at which the probability of a state being occupied is equal to one half. For a crystal no longer in equilibrium, the occupation probability of any state of the conduction band is given by

$$f_c = \frac{1}{1 + \exp [E(k) - F_c]/kT}$$

where F_c is the "quasi Fermi level" for the electrons of the conduction band. A "quasi Fermi level" F_v may also be defined for the holes of the valence band and at equilibrium, clearly,

$$F_c = F_v = F_0$$

If W_v^c is the probability per unit time of a direct radiative transition between the valence and conduction bands, then in a radiation field consisting of photons of energy $h\nu$ and density $P(\nu)$, the number of quanta absorbed per unit time is,

$$N_a = A \,.\, W_v^c f_v(k_i)[1 - f_c(k_j)]P(\nu)$$

The number of quanta emitted per unit time by stimulated emission is given by

$$N_e = A \,.\, W_c^v[1 - f_v(k_i)]f_c(k_j)P(\nu)$$

so for

$$N_e > N_a$$

taking $W_v^c = W_c^v$, then

$$f_c(k_j)[1 - f_v(k_i)] > f_v(k_i)[1 - f_c(k_j)]$$

$$\therefore \quad \exp\frac{(F_c - F_v)}{kT} > \exp\frac{[E_c(k_j) - E_v(k_i)]}{kT}$$

and since

$$E_c(k_j) - E_v(k_i) = h\nu$$

then

$$F_c - F_v > h\nu$$

Bernard and Duraffourg showed that a similar condition exists for a transition between the conduction band and an impurity level in the forbidden energy gap. They showed too that provided a "quasi Fermi level" may be defined for groups of different types of electronic states in a semiconductor, the necessary condition for electromagnetic waves to be amplified is given by the above relationship. The different types of electronic states include conduction band, valence band and impurity levels.

Figure 5.1b shows the situation for an inversion of population at 0°K. The valence band is empty of electrons down to level F_v and the conduction band filled up to F_c. Stimulated emission can occur, and the resulting photons have energy $h\nu$ where $E_G < h\nu < (F_c - F_v)$. At a finite temperature the sharp distinction between filled and unfilled states will not occur; the distributions of the carriers will be smeared out as shown in Fig. 5.1c. This is why Bernard and Duraffourg found it necessary to calculate the occupation probability of any energy state using the Fermi–Dirac function.

Two types of transition may occur, the direct and the indirect. For the direct process to be possible the energy band structure must be such that the lowest conduction band minimum and the highest valence band maximum have the same wave vector in the Brillouin zone (as in Fig. 5.1). When the extrema are at different wave vectors indirect transitions occur because of the need for a lattice vibration, or an impurity, to take up the change in the wave-vector. In general it would be anticipated that direct transitions would have more gain associated with them.

For a direct transition $h\nu = E_G$, where E_G is the width of the energy gap, the condition $F_c - F_v > h\nu$ requires that the number of carriers injected be so large that both the electron and hole populations are degenerate, or nearly so. If an indirect transition occurs, involving the emission of a phonon, then $h\nu$ is much less than the energy gap and the laser pump power is greatly reduced. Here,

$$h\nu = E_G - \hbar\omega$$

where $\hbar\omega$ is the energy of the phonon.

The role of impurities in promoting conduction of a current in a semiconductor is an important one. A donor impurity is one that has more valence electrons than the host atom for which it substitutes. These excess electrons can be donated to the conduction band. The lowest state for these additional electrons will, in general, be within the forbidden gap fairly close to the conduction band. At room temperature some of the electrons will be in the conduction band. An acceptor impurity has fewer electrons than the atom for which it substitutes. It takes electrons from the valence band to make up for this and leaves holes behind. The state to which the valence electrons must be promoted also lies in the forbidden gap, usually close to the valence band.

An *n*-type semiconductor has more donors than acceptors and a *p*-type semiconductor has more acceptors than donors. Radiative recombination may occur between (a) a conduction band free

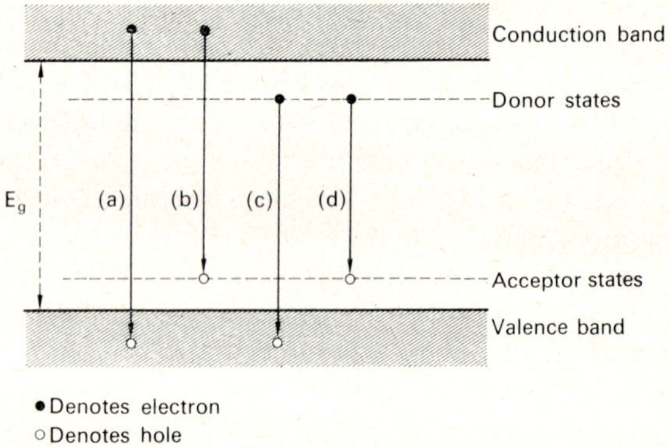

FIG. 5.2. Four possible radiative recombination processes between holes and electrons.

electron and a valence band hole, (b) between a free electron and a hole in the acceptor state just above the valence band, (c) between a free hole in the valence band and an electron in the donor state just below the conduction band, and (d) between an electron in a donor state and a hole in an acceptor state (see Fig. 5.2).

An efficient semiconductor laser material must combine a fast radiative recombination path for the holes and electrons, and a suitable structure for introducing electrons into excited states.

Three basic techniques exist to do this; by injecting holes and electrons into an insulating region; by injecting electrons into *p*-type material, and by injecting holes into *n*-type material.

The most effective system for injecting carriers into a semiconducting material, to such an extent that population inversion occurs, is the *p–n* junction. Figure 5.3a shows an energy diagram of a *p–n* junction. In the *n*-type semiconductor there are enough electrons added by impurities to fill the conduction band up to Fermi level *F*. In the *p*-type, acceptor impurities have been added which

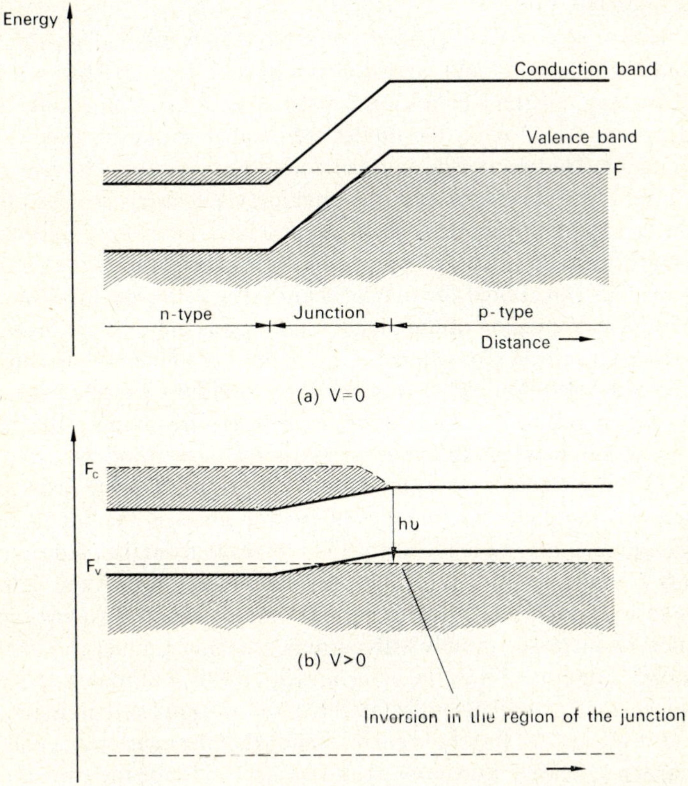

FIG. 5.3. Population inversion in a *p–n* junction.

add holes down to energy F. At zero applied voltage, electrons flow from the n- to the p-sides until an electrical potential barrier V_B is built up which prevents further flow of current. When a voltage is applied which raises the n- relative to the p-side, as shown in Fig. 5.3b, then electrons can flow to the p-side where they make transitions to empty states in the valence band and emit photons of energy, approximately E_G. In actual junctions there are impurity levels in the gap near the band limits and the photon energy will be less than E_G. For a high enough applied voltage, there can be a region in the vicinity of the junction where the population is inverted.

Hall *et al.* (1962) (Reprinted Paper No. 11), Nathan *et al.* (1962) and Quist *et al.* (1962) virtually simultaneously, published work describing coherent light emission from GaAs p–n junctions. In the case of Hall *et al.* the diodes were cubes with each edge 0·4 mm long. The junction lay in a horizontal plane through the centre. Current was passed through the junction via contacts attached to the top and bottom faces. The front and back faces were polished parallel to each other and perpendicular to the plane of the junction. This constituted the optical cavity. The active medium was, of course, the region of the junction and it was only a fraction of a wavelength thick. Consequently the pattern of the emitted radiation did not follow the characteristic distributions expected from the theory of Fox and Li. Interference fringes were obtained due to interaction between different parts of the wavefront and their appearance showed that a definite phase relationship existed between the light emitted at different points in the junction. Measurements were made of the light intensity from the diode as a function of current. It was found that below 5000 A/cm^2 the light intensity varied linearly with current density. At 8500 A/cm^2 the intensity increased rapidly with current, reaching a value, at 20,000 A/cm^2, about ten times the value anticipated by extrapolating the low current plot. This indicated the onset of stimulated emission, and it was at this threshold current value that the interference phenomena appeared. Furthermore the spectral width of the emission below threshold was 125 Å and above threshold it decreased suddenly to 15 Å. Similar behaviour was observed by Nathan *et al.* and was taken, correctly, to be evidence of the presence of stimulated emission.

An alternative pumping method is electron beam pumping. A beam of high energy electrons, at least 20 kV, is directed at a semiconductor. These electrons penetrate several microns into the material and lose a large fraction of their energy by creating many low energy electron-hole pairs. Each incident electron creates $\sim 10^4$ electrons and holes. These decay to the conduction and valence bands respectively and create an inverted population. Optical pumping has also been used. Photons of greater energy than the energy gap excite the semiconductor and the situation is then identical to electron beam pumping.

In the absence of inversion, light emission by the recombination of electrons and holes may still occur, and semiconducting light sources are now available over a wide range of wavelengths. It can be difficult for the photons to escape being re-absorbed by the junction, or semiconductor, material. When population inversion does occur, the gain of the system at threshold just equals the losses due to re-absorption plus those due to the imperfectly reflecting end faces and true laser oscillation is able to commence. A useful review of semiconductor lasers is contained in Lengyel's book (1966).

CHAPTER 6

A More Complex View of the Laser

So FAR, emphasis has been laid on the essential features of the various types of laser, and on outlining a simple theoretical basis for laser action. Although experimental verification of Fox and Li type modes has been quoted, Bennett (1962) (Reprinted Paper No. 12) has shown that the ideas expressed about mode spacing are, in fact, too simple. Similarly the theories of Schawlow and Townes and of Prokhorov are incapable of accounting for a great many experimental observations. They are unable to account for the way in which several modes resonate simultaneously at different frequencies across the gain profile of the atomic resonance. Nor can the degree of inversion at the centre of an atomic resonance be satisfactorily associated with the resulting output intensity of a laser. The more that has been found out about the properties of laser systems the greater has become the need for an all-embracing theory. Many theories have been put forward and, it should be made clear, none are totally satisfactory. The semi-classical theory of Lamb (1964) has proved to be the most useful and, although the paper is not reprinted here, a brief outline of the work will be given, if only to show that laser theory is by no means as simple as the preceding chapters of this book may suggest. Fully quantum-mechanical treatments of the problem such as the one by Scully and Lamb (1967) will no doubt prove to be more rigorous and more useful, but they are intrinsically more complex and outside the scope of this book.

It was noted in Chapter 4 that Herriott had experimentally verified the idea that a number of axial modes, separated by $\Delta\nu = c/2d$ Hz, might be expected to resonate within the Doppler-broadened linewidth of a given atomic transition. He had examined the signal from a photo-multiplier for beats and found a component at $c/2d$. Other beat notes were accounted for by the Fox and Li prediction that off-axis modes also existed which were slightly displaced in frequency from the axial modes. Bennett repeated the experiment using the $1{\cdot}15\ \mu$ line of He–Ne. He found that the beat frequency between axial modes was slightly less than $c/2d$; furthermore, the frequency of the $\sim c/2d$ beat increased with increasing power and, as the power level increased still more a power-dependent splitting of the beat was encountered. In particular it was found that at low powers a single beat was observed at just less than $c/2d$ and as the laser power was increased the $c/2d$ beat split by $\sim$20 kHz, and the $2(c/2d)$ beat appeared. When the power was increased once more, the $c/2d$ beat became triple and the $2(c/2d)$ beat double, and so on.

These observations may be accounted for on the assumption that some non-linear frequency-dependent pulling mechanism exists whereby the frequency spacings between adjacent resonant modes are not identical. In this case, if the number of beats in the region of $\sim c/2d$ is m, the number of axial modes oscillating must be $m + 1$. Bennett assumed that no coupling effects existed between simultaneously oscillating modes through time-dependent nonlinearities in the medium, but that the introduction of the amplifying medium changes the refractive index of the system and so alters the single-pass phase shift from that which occurred in the passive case.

He assumed that in the absence of gain the mode frequency of the passive cavity was ν_{0c}. When gain occurred due to the presence of an active medium the oscillation frequency shifts to $\bar{\nu}$. Since the condition for a mode to exist is that the net phase shift per pass is unchanged,

$$\frac{\partial \phi_c}{\partial \nu}(\bar{\nu} - \nu_{0c}) + \Delta\phi_m(\bar{\nu}) = 0 \tag{1}$$

where $\Delta\phi_m(\bar{\nu})$ is the change in single-pass phase shift at the frequency of oscillation due to the fractional gain per pass, while

$\partial\phi_c/\partial\nu$ is the rate of change of phase with frequency. Now the single pass phase shift for the cavity in the absence of gain is,

$$\phi_c = \frac{2\pi}{\lambda} d = \frac{2\pi\, d}{c} \nu \tag{2}$$

But

$$\Delta\nu_c = \frac{\nu}{Q} = \text{bandwidth of cavity resonance}$$

and Q may be defined as

$$Q = \frac{\phi_c}{f}$$

where f is the loss per pass.

$$\therefore \quad \Delta\nu_c = \frac{c}{2\pi\, d} f \tag{3}$$

From (2) and (3)

$$\frac{\partial\phi_c}{\partial\nu} = \frac{2\pi\, d}{c} = \frac{f}{\Delta\nu_c}$$

and so, from (1),

$$\bar{\nu} = \nu_{0c} - \frac{\Delta\nu_c}{f} \Delta\phi_m(\bar{\nu})$$

and $\nu_{0c} - \bar{\nu}$ represents the degree of "pulling" of the mode due to the presence of gain in the system. The term $\Delta\phi_m(\bar{\nu})$ is a function of the gain per pass; and, since gain varies with frequency across the line profile of the transition, so too does $\Delta\phi_m(\bar{\nu})$. In general it is zero at the line centre, negative for lower frequencies and positive for higher frequencies. This is due to anomalous dispersion in the region of an absorption line. Bennett shows, using the Kramers–Kronig relationships (see Ditchburn, 1963), that to a good approximation for a Gaussian line shape, as for a Doppler-broadened transition,

$$\Delta\phi_m(\bar{\nu}) = -0{\cdot}28\, g(\nu) \exp\left[-\left(\frac{\nu_m - \nu}{0{\cdot}6\Delta\nu_m}\right)^2\right] \sin\left(\frac{\nu_m - \nu}{0{\cdot}3\Delta\nu_m}\right)$$

where $g(\nu)$ is the gain at frequency ν.

This "mode pulling" accounts very well for the observed axial mode beat of just less than $c/2d$ and it accounts, too, for the splitting of the beat as other axial modes start to oscillate as the input power is increased. It does not, however, account for the power ting of the beat as other axial modes start to oscillate with increasing input power. It does not, however, account for the power dependence of the actual frequencies between the beat components at $\sim c/2d$ and $\sim 2(c/2d)$. Bennett accounted for this by considering the effect of "hole burning".

For an inhomogeneously broadened line, such as a Doppler-broadened atomic transition in a gas, the reduction of the gain at a certain frequency $\bar{\nu}$ which saturates at $g(\bar{\nu}) = f$ (see Fig. 4.2) does not imply a reduction of the gain at other frequencies. At frequencies corresponding to cavity resonances well removed from the "hole" at $\bar{\nu}$, the phase shift continues to increase as the number of upper state atoms increases. However, the true phase shift at that frequency can be found by subtracting from the results of the previous analysis the effect of the "hole" at $\bar{\nu}$. The presence of a "hole" at one resonance reduces the degree of "pulling" at another; so, although the two oscillations involve different atoms, there is a large first order interaction between the two.

Bennett showed that the probability for stimulated emission is a Lorentzian function of the frequency separation between the mode of oscillation and the centre frequency of the atoms involved. The width of the Lorentzian at half-maximum turns out to be a function of the power in the mode, and so of the gain at resonance. The total phase shift introduced at frequency ν_j by N Lorentzian "holes" at frequency ν_i is then given by

$$\Delta\phi_H(j) \approx \sum_{\substack{i=1 \\ i \neq j}}^{N} \left(\frac{g(i) - f}{4}\right)\left(\frac{H_i}{\nu_i - \nu_j}\right)$$

The oscillation frequency for the *j*th cavity resonance, taking into account both "mode pulling" and "hole burning", is

$$\bar{\nu}_j = \nu_j - \frac{\Delta\nu_c}{f}\Delta\phi_m(\nu_j) - \frac{\Delta\nu_c}{4}\sum_{\substack{i=1 \\ i\neq j}}^{N}\left(\frac{g(i)}{f} - 1\right)\left(\frac{H_i}{\nu_i - \nu_j}\right)$$

where H_i is the width of the hole of frequency ν_i. In conclusion Bennett demonstrated that the observed beat splittings, and their dependence on power, were in good agreement with this theory.

Bennett did not at the time realise, although he did subsequently (1964), that, on the basis of this model, there would be two "holes' burned for each resonance. Consider the standing wave in the cavity to be made up of a pair of running waves, one in each direction. The running wave in the $+z$ direction will interact primarily with atoms whose Doppler-shifted transition frequencies, ν_0, arise from a velocity component $+v$, where

$$\nu_0 = \left(1 + \frac{v}{c}\right)\nu_m$$

That in the $-z$ direction interacts mainly with atoms moving with velocity $-v$ at a frequency ν', where

$$\nu' = \left(1 - \frac{v}{c}\right)\nu_m$$

Hence, two "holes" are burned in the profile; one at ν_0 and the other $\nu' = 2\nu_m - \nu_0$. Only when $\nu_0 = \nu_m$ will there be a single hole. The energy emitted by a laser is clearly proportional to the number of atoms "burned out" by stimulated emission, that is to the area of the "hole" created. It is possible, by for instance making the mode separation greater than the Doppler width (i.e. $c/2d > \Delta\nu_m$), to make a single mode laser. If the oscillation frequency is varied linearly across the gain profile then, due to the two "holes" being coincident at ν_m, there will be a dip in the output intensity at that frequency. This is known as the Lamb Dip since Lamb (1964), from a completely different theoretical standpoint, predicted its existence.

Lamb's theory is a semi-classical one; the radiation field is not quantized but is treated classically in its interaction with a number of two level atoms. The basis of the theory is the fact that the field pattern in the laser cavity is sustained by stimulated emission from the active medium, but the behaviour of the medium itself is modified by the existence of the field. The field in the cavity produces a macroscopic electric polarization of the medium and this polarization acts as a source for the field itself. A self-consistent field theory results since the polarization and the field may be connected by Maxwell's equations.

An appropriate form of the equation connecting the polarization of the medium $\boldsymbol{P}$ (r,t) and the field in the cavity $\boldsymbol{E}$ (r,t) is,

$$\nabla x \nabla x \boldsymbol{E} + \mu_0 \sigma \frac{\partial \boldsymbol{E}}{\partial t} + \mu_0 \varepsilon_0 \frac{\partial^2 \boldsymbol{E}}{\partial t^2} = -\mu_0 \frac{\partial^2 \boldsymbol{P}}{\partial t^2}$$

where diffraction and reflection losses are represented by the conductivity term σ. Assuming that the only important variations of E and P are along the length of the cavity, in direction z, the above equation becomes

$$-\frac{\partial^2 E}{\partial z^2} + \mu_0 \sigma \frac{\partial E}{\partial t} + \mu_0 \varepsilon_0 \frac{\partial^2 E}{\partial t^2} = -\mu_0 \frac{\partial^2 P}{\partial t^2} \qquad (1)$$

where

$$E = E(z,t) \quad \text{and} \quad P = P(z,t)$$

Now the loss less normal modes of the system may be given by eigenfunctions

$$E_n = e^{i\Omega_n t}\, U_n(z)$$

where

$$U_n(z) = \sin\left(\frac{n\pi z}{d}\right)$$

and $\Omega_n = \dfrac{\pi n c}{d}$ the resonant cavity frequencies. If the loss term in equation (1) is treated as a perturbation of the normal mode structure then, in the presence of the polarization $P(z,t)$, quasi-stationary forced oscillations of the electric field can be expressed as

$$E(z,t) = \sum A_n(t)U_n(z) \tag{2}$$

Substitute (2) into (1),

$$-\sum A_n(t)\frac{\partial^2 U_n(z)}{dz^2} + \sum \mu_0 \sigma \frac{\partial A_n(t)}{\partial t} U_n(z)$$
$$+ \sum \mu_0 \varepsilon_0 \frac{\partial^2 A_n(t)}{\partial t^2} U_n(z) = -\mu_0 \frac{\partial^2 P}{\partial t^2}$$

Now,
$$U_n(z) = \sin\left(\frac{n\pi z}{d}\right)$$

$$\therefore \quad \frac{\partial^2 U_n(z)}{\partial z^2} = \frac{-n^2\pi^2}{d^2}\sin\left(\frac{n\pi z}{d}\right) = -\frac{\Omega_n^2}{c^2}U_n(z)$$

$$\therefore \quad \sum A_n(t)\frac{\Omega_n^2}{c^2}U_n(z) + \sum \mu_0 \sigma \frac{\partial A_n(t)}{\partial t} U_n(z)$$
$$+ \sum \mu_0 \varepsilon_0 \frac{\partial^2 A_n}{\partial t^2} U_n(z) = -\mu_0 \frac{\partial^2 P}{\partial t^2}$$

Multiply through by $U_m(z)$, integrate over z and use the orthogonality relationship for eigenfunctions,

$$\therefore \quad A_n(t)\frac{\Omega_n^2}{c^2} + \mu_0 \sigma \frac{\partial A_n(t)}{\partial t} + \mu_0 \varepsilon_0 \frac{\partial^2 A_n}{\partial t^2}$$
$$= -\mu_0 \frac{\partial^2}{\partial t^2}\int P(z,t)U_n(z)\,dz$$

Multiply through by $\varepsilon_0 c^2$,

$$\varepsilon_0 \Omega_n^2 A_n(t) + \sigma \frac{dA_n(t)}{dt} + \varepsilon_0 \frac{d^2 A_n}{dt^2} = \frac{-d^2 P_n(t)}{dt^2} \tag{3}$$

where

$$P_n(t) = \frac{2}{d}\int_0^d P(z,t)U_n(z)\,dz$$

which is the spatial Fourier component of $P(z,t)$.

Now assuming

$$\nu \approx \Omega_n$$

and

$$P_n \approx P_0 \sin \nu t$$

so that

$$\ddot{P}_n \approx -\nu^2 P_n$$

substituting into (3) and putting

$$\sigma = \frac{\varepsilon_0 \nu}{Q}$$

where ν is the damped frequency of free oscillation,

$$\frac{d^2A_n}{dt^2} + \left(\frac{\nu}{Q}\right)\frac{dA_n}{dt} + \Omega_n^2 A_n = \frac{\nu^2}{\varepsilon_0} P_n \tag{4}$$

which is the equation of a forced harmonic oscillator.

Since $\Delta\nu_c$, the cavity bandwidth, is narrow compared with the mode separation $\sim c/2d$, Fourier components of $A_n(t)$ and $P_n(t)$ at frequencies far from resonance may be neglected,

$$\therefore \quad A_n(t) = E_n(t) \cos\left[\nu_n t + \phi_n(t)\right] \tag{5}$$

and

$$P_n(t) = C_n(t) \cos(\nu_n t + \phi_n(t)) + S_n(t) \sin(\nu_n t + \phi_n(t)) \tag{6}$$

Put (5) and (6) into (4) and equate coefficients,

$$(\nu_n + \dot{\phi} - \Omega_n)E_n = -\tfrac{1}{2}\left(\frac{\nu}{\varepsilon_0}\right) C_n$$

and

$$\dot{E}_n + \tfrac{1}{2}\left(\frac{\nu}{Q_n}\right) E_n = -\tfrac{1}{2}\left(\frac{\nu}{\varepsilon_0}\right) S_n$$

These equations determine the amplitudes, frequencies and phases of the optical resonances of the system. The problem then remains to determine the coefficients C_n and S_n by finding P_n in terms of E_n. Lamb did this by considering the interaction of the electromagnetic wave in the cavity with the individual atoms making up the polarized medium. He solved the problem to both first and third order perturbation theory, taking into account the fact that the atoms have a velocity distribution and that their time and place of interaction with the electromagnetic field differs from the time and place of their excitation.

Many important results have been drawn from the theory. Terms equivalent to Bennett's "mode pulling" occur; "mode pushing" is found and this may be equated, loosely speaking, to "hole burning". The "Lamb Dip" is predicted and has since been observed; equations describing the interaction of several modes oscillating at the same time are given, and many subsequently confirmed phenomena described. Although attempts have been made at fully quantum mechanical treatments of the problem, none, so far, have been able to describe such a wealth of phenomenological detail as this theory of Lamb's.

The field of laser physics has made great strides in the last few years; lasers have proved an interesting problem to physicists and technologists. The preceding chapters are an introduction to this still fast growing field, and some feeling for the enormous range of experiments and ideas concerning lasers may be achieved by reading the more advanced books that have been cited.

REFERENCES

AIGRAIN, P. (1958) Unpublished lecture at the International Conference on Solid State Physics in Electronics and Telecommunications, Brussels, 1958.

ALLEN, L. and JONES, D. G. C. (1967) *Principles of Gas Lasers*, Butterworths, London.

BENNETT, Jr., W. R. (1962) *Phys. Rev.* **126,** 580.

BENNETT, Jr., W. R. (1964) *Quantum Electronics III* (ed. Grivet and Bloembergen), Dunod, Paris.

BERNARD, M. G. A. and DURAFFOURG, G. (1961) *Physica Status Solidi* **1,** 699.

BORN, M. and WOLF, E. (1964) *Principles of Optics*, Pergamon Press, Oxford.

BOYD, G. D. and GORDON, J. P. (1961) *Bell Syst. Tech. Journ.* **40,** 489.

BOYD, G. D. and KOGELNIK, H. (1962) *Bell Syst. Tech. Journ.* **41,** 1347.

COLLIN, R. E. (1960) *Field Theory of Guided Waves*, McGraw-Hill, New York.

COLLINS, Jr., S. A. (1963) *J.O.S.A.* **53,** 1339.

DITCHBURN, R. W. (1963) *Light*, Blackie, London, 1963.

FOX, A. G. and LI, T. (1961) *Bell Syst. Tech. Journ.* **40**, 453.

GARRETT, G. (1961) *Gas Lasers*, McGraw-Hill, New York.

GORDON, J. R., ZEIGER, H. Z. and TOWNES, C. H. (1955) *Phys. Rev.* **99,** 1264.

HALL, R. N., FENNER, G. E., KINGSLEY, J. D., SOLTYS, T. J. and CARLSON, R. O. (1962) *Phys. Rev. Letts.* **9,** 366.

HERRIOTT, D. R. (1962) *J.O.S.A.* **52,** 31.

HOLSTEIN, T. (1947) *Phys. Rev.* **72,** 212.

JACOBS, S., GOULD, G. and RABINOWITZ, P. (1961) *Phys. Rev. Letts.* **7,** 415.

JAVAN, A. (1959) *Phys. Rev. Letts.* **3,** 88.

JAVAN, A., BENNETT, Jr., W. R. and HERRIOT, D. R. (1961) *Phys. Rev. Letts.* **6,** 106.

KAMINSKY, A. A. and OSIKO, B. B. (1965) *Inorganic Materials* **1,** 2049.

KISS, Z. J. and PRESSLEY, R. (1966) *J. App. Opt.* **5,** 1474.

KOGELNIK, H. and LI, T. (1966) *App. Opt.* **5,** 1550.

LAMB, Jr., W. E. (1964) *Phys. Rev.* **134,** 1429.

LAMB, Jr., W. E. (1965) *Quantum Optics and Electronics* (ed. DE WITT), Gordon & Breach, New York.

LENGYEL, B. A. (1966) *Introduction to Laser Physics*, Wiley.

MAIMAN, T. H. (1960a) *Phys. Rev. Letts.* **4,** 564.

MAIMAN, T. H. (1960b) *Nature* **187,** 493.

MAIMAN, T. H. (1961) *Phys. Rev.* **123,** 1145.

MITCHELL, A. C. G. and ZEMANSKY, M. W. (1934) Cambridge University Press.

MOORE, C. E. (1949) *Atomic Energy Levels*, vol. I. N.B.S., Washington, Circular No. 467.

NATHAN, M. I., DUMKE, W. P., BURNS, G., DILL, F. H. and LASHER, G. (1962) *App. Phys. Letts.* **1,** 62.
NELSON, D. F. and BOYLE, W. S. (1962) *App. Opt.* **1,** 181.
PROKHOROV, A. M. (1958) *Soviet-Phys.-J.E.T.P.* **7,** 1140.
QUIST, T. M., REDIKER, R. H., KEYES, R. J., KRAG, W. E., LAX, B., MCWHORTER, A. L. and ZEIGER, H. J. (1962) *App. Phys. Letts.* **1,** 91.
SCHAWLOW, A. and TOWNES, C. H. (1958) *Phys. Rev.* **112,** 1940.
SCULLY, M. O. and LAMB, Jr., W. E. (1967) *Phys. Rev.* **159,** 208.
SMITH, W. V. and SOROKIN, P. P. (1966) *The Laser*, McGraw-Hill, New York.
SOROKIN, P. P. and STEVENSON, M. J. (1960) *Phys. Rev. Letts.* **5,** 557.
VAN DER POL, B. (1934) *Proc. I.R.E.* **22,** 1051.
YARIV, A. and GORDON, J. P. (1963) *Proc. I.E.E.E.* **51,** 4.

PART II
REPRINTED PAPERS

1

Infrared and Optical Lases*

A. L. SCHAWLOW AND C. H. TOWNES†

Bell Telephone Laboratories, Murray Hill, New Jersey

Summary

The extension of maser techniques to the infrared and optical region is considered. It is shown that by using a resonant cavity of centimeter dimensions, having many resonant modes, maser oscillation at these wave-lengths can be achieved by pumping with reasonable amounts of incoherent light. For wavelengths much shorter than those of the ultraviolet region, maser-type amplification appears to be quite impractical. Although use of a multimode cavity is suggested, a single mode may be selected by making only the end walls highly reflecting, and defining a suitably small angular aperture. Then extremely monochromatic and coherent light is produced. The design principles are illustrated by reference to a system using potassium vapor.

Introduction

Amplifiers and oscillators using atomic and molecular processes, as do the various varieties of masers,[1-4] may in principle be extended far beyond the range of frequencies which have been generated electronically, and into the infrared, the optical region, or beyond. Such techniques give the attractive promise of coherent amplification at these high frequencies and of generation of very monochromatic radiation. In the infrared region in particular, the generation of reasonably intense and monochromatic radiation would allow the possibility of spectroscopy at very much higher

* *Phys. Rev.* **112,** 1940 (1958).

† Permanent address: Columbia University, New York, New York.

[1] Gordon, Zeiger and Townes, *Phys. Rev.* **99,** 1264 (1955).

[2] Combrisson, Honig, and Townes, *Compt. rend.* **242,** 2451 (1956),

[3] N. Bloembergen, *Phys. Rev.* **104,** 329 (1956).

[4] E. Allais, *Compt. rend.* **245,** 157 (1957).

resolution than is now possible. As one attempts to extend maser operation towards very short wavelengths, a number of new aspects and problems arise, which require a quantitative reorientation of theoretical discussions and considerable modification of the experimental techniques used. Our purpose is to discuss theoretical aspects of maser-like devices for wavelengths considerably shorter than one centimeter, to examine the short-wavelength limit for practical devices of this type, and to outline design considerations for an example of a maser oscillator for producing radiation in the infrared region. In the general discussion, roughly reasonable values of design parameters will be used. They will be justified later by more detailed examination of one particular atomic system.

Characteristics of Masers for Microwave Frequencies

For comparison, we shall consider first the characteristics of masers operating in the normal microwave range. Here an unstable ensemble of atomic or molecular systems is introduced into a cavity which would normally have one resonant mode near the frequency which corresponds to radiative transitions of these systems. In some cases, such as ensemble may be located in a wave guide rather than in a cavity but again there would be characteristically one or a very few modes of propagation allowed by the wave guide in the frequency range of interest. The condition of oscillation for n atomic systems excited with random phase and located in a cavity of appropriate frequency may be written (see references 1 and 2)

$$n \geq hV\Delta\nu/(4\pi\mu^2 Q_c), \tag{1}$$

where n is more precisely the difference n_1-n_2 in number of systems in the upper and lower states, V is the volume of the cavity, $\Delta\nu$ is the half-width of the atomic resonance at half-maximum intensity, assuming a Lorentzian line shape, μ is the matrix element involved in the transition, and Q_c is the quality factor of the cavity.

The energy emitted by such a maser oscillator is usually in an extremely monochromatic wave, since the energy produced by stimulated emission is very much larger than that due to spontaneous emission or to the normal background of thermal radiation. The frequency range over which appreciable energy is distributed is given approximately by[1]

$$\delta\nu = 4\pi kT(\Delta\nu)^2/P, \tag{2}$$

where $\Delta\nu$ is the half-width at half-maximum of the resonant response of a single atomic system, P is the total power emitted, k is Boltzmann's constant, and T the absolute temperature of the cavity walls and wave guide. Since in all maser oscillators at microwave frequencies which have so far been considered, $P \gg kT\Delta\nu$, the radiation is largely emitted over a region very much smaller than $\Delta\nu$, or $\delta\nu \ll \Delta\nu$.

As amplifiers of microwave or radio-frequency energy, masers have the capability of very high sensitivity, approaching in the limit the possibility of detecting one or a few quanta. This corresponds to a noise temperature of $h\nu/k$, which for microwave frequencies is of the order of 1 °K.

Use of Multimode Cavities at High Frequencies

Consider now some of the modifications necessary to operate a maser at frequencies as high as that of infrared radiation. To maintain a single isolated mode in a cavity at infrared frequencies, the linear dimension of the cavity would need to be of the order of one wavelength which, at least in the higher frequency part of the infrared spectrum, would be too small to be practical. Hence, one is led to consider cavities which are large compared to a wavelength, and which may support a large number of modes within the frequency range of interest. For very short wavelengths, it is perhaps more usual to consider a plane wave reflected many times from the walls of such a cavity, rather than the field of a standing wave which would correspond to a mode.

The condition for oscillation may be obtained by requiring that the power produced by stimulated emission is as great as that lost to the cavity walls or other types of absorption, that is,

$$\left(\frac{\mu' E}{\hbar}\right)^2 \frac{h\nu n}{4\pi\Delta\nu} \geq \frac{E^2}{8\pi}\frac{V}{t}, \tag{3}$$

where μ' is the matrix element for the emissive transition, E^2 is the mean square of the electric field. (For a multiresonant cavity, E^2 may be considered identical in all parts of the cavity.) n is the excess number of atoms in the upper state over those in the lower state, V is the volume of the cavity, t is the time constant for the rate of decay of the energy, $\Delta\nu$ is the half-width of the resonance at half maximum intensity, if a Lorentzian shape is assumed. The decay time t may be writen as $Q/2\pi\nu$, but is perhaps more naturally expressed in terms of the reflection coefficient α of the cavity walls.

$$t = 6V/(1-\alpha)Ac, \tag{4}$$

where A is the wall area and c the velocity of light. For a cube of dimension L, $t = L/(1-\alpha)c$. The condition for oscillation from (3) is then

$$n \geq \frac{3hV}{8\pi^2\mu^2 t}\frac{\Delta\nu}{\nu}, \tag{5}$$

or

$$n \geq \frac{\Delta\nu}{\nu}\frac{h(1-\alpha)Ac}{16\pi^2\mu^2}. \tag{6}$$

Here μ'^2 has been replaced by $\mu^2/3$, since μ'^2 is the square of the matrix element for the transition which, when averaged over all orientations of the system, is just one-third of the quantity μ^2 which is usually taken as the square of the matrix element.

In a gas at low pressure, most infrared or optical transitions will have a width $\Delta\nu$ determined by Doppler effects. Then the resonance half-width at half-maximum intensity is

$$\Delta\nu = \frac{\nu}{c}\left(\frac{2kT}{m}\ln 2\right)^{\frac{1}{2}}, \tag{7}$$

where m is the molecular mass, k is Boltzmann's constant, and T the temperature. Because of the Gaussian line shape in this case, expression (6) becomes

$$n \geq \frac{\Delta v}{v} \frac{h(1-\alpha)Ac}{16\pi^2\mu^2(\pi \ln 2)^{\frac{1}{2}}}, \tag{8}$$

or

$$n \geq \frac{h(1-\alpha)A}{16\pi^2\mu^2}\left(\frac{2kT}{\pi m}\right)^{\frac{1}{2}}. \tag{9}$$

It may be noted that expression (9) for the number of excited systems required for oscillation is independent of the frequency. Furthermore, this number n is not impractically large. Assuming the cavity is a cube of 1 cm dimension and that $\alpha = 0.98$, $\mu = 5 \times 10^{-18}$ esu, $T = 400°$K, and $m = 100$ amu, one obtains $n = 5 \times 10^8$.

The condition for oscillation, indicated in (5), may be conveniently related to the lifetime τ of the state due to spontaneous emission of radiation by a transition between the two levels in question. This lifetime is given, by well-known theory, as

$$\tau = 3hc^3/(64\pi^4 v^3 \mu^2). \tag{10}$$

Now the rate of stimulated emission due to a single quantum in a single mode is just equal to the rate of spontaneous emission into the same single mode. Hence, $1/\tau$ is this rate multiplied by the number of modes p which are effective in producing spontaneous emission. Assuming a single quantum present in a mode at the resonant frequency, the condition for instability can then be written

$$nhv/p\tau \geq hv/t_1,$$

or

$$n \geq p\tau/t. \tag{11}$$

This gives a simple expression which may sometimes be useful, and which is equivalent to (5), since

$$p = \int p(\nu) \frac{(\Delta\nu)^2 \, d\nu}{(\nu - \nu_0)^2 + (\Delta\nu)^2}, \tag{12}$$

where $p(\nu)d\nu$ is the number of modes between ν and $\nu + d\nu$, which is well known to be

$$p(\nu) \, d\nu = 8\pi\nu^2 V \, d\nu/c^3 . \tag{13}$$

From (12) and (13), one obtains for a Lorentzian line shape,

$$p = 8\pi^2\nu^2 V\Delta\nu/c^3. \tag{14}$$

Or, for a line broadened by Doppler effects, the corresponding number of effective modes is

$$p = 8\pi^2\nu^2 V \, \Delta\nu/(\pi \ln 2)^{\frac{1}{2}}c^3 . \tag{15}$$

If r and p are inserted into (11) from expressions (10) and (14), respectively, it becomes identical with (5), as one must expect.

The mimimum power which must be supplied in order to maintain n systems in excited states is

$$P = nh\nu/\tau = ph\nu/t . \tag{16}$$

This expression is independent of the lifetime or matrix element. However, if there are alternate modes of decay of each system, as by collisions or other transitions, the necessary power may be larger than that given by (16) and dependent on details of the system involved. Furthermore, some quantum of higher frequency than that emitted will normally be required to excite the system, which will increase the power somewhat above the value given by (16). Assuming the case considered above, i.e., a cube of 1-cm dimension with $\alpha = 0.98$, $\lambda = 10^4$ Å, and broadening due to Doppler effect, (16) gives $P = 0.8 \times 10^{-3}$ watt. Supply of this much power in a spectral line does not seem to be extremely difficult.

The power generated in the coherent oscillation of the maser may be extremely small, if the condition of instability is fulfilled in a very marginal way, and hence can be much less than the total power, which would be of the order of 10^{-3} watt, radiated spontaneously. However, if the number of excited systems exceeds the critical number appreciably, e.g., by a factor of two, then the power of stimulated radiation is given roughly by $h\nu$ times the rate at which excited systems are supplied, assuming the excitation is not lost by some process not yet considered, such as by collisions. The electromagnetic field then builds up so that the stimulated emission may be appreciably greater than the total spontaneous emission. For values even slightly above the critical number, the stimulated power is of the order of the power $nh\nu/\tau$ supplied, or hence of the order of one milliwatt under the conditions assumed above.

The most obvious and apparently most convenient method for supplying excited atoms is excitation at a higher frequency, as in optical pumping or a three-level maser system. The power supplied must, of course, be appreciably greater than the emitted power in expression (9). There is no requirement that the pumping frequency be much higher than the frequency emitted, as long as the difference in frequency is much greater than kT/h, which can assure the possibility of negative temperatures, Since, for the high frequencies required, an incoherent source of pumping power must be used, a desirable operating frequency would be near the point where the maximum number of quanta are emitted by a given transition from a discharge or some other source of high effective temperature. This maximum will occur somewhere near the maximum of the blackbody radiation at the effective temperature of such a source, or hence in the visible or ultraviolet region. The number of quanta required per second would probably be about one order of magnitude greater than the number emitted at the oscillating frequency, so that the input power required would be about ten times the output given by (16), or 10 milliwatts. This amount of energy in an individual spectroscopic line is, fortunately, obtainable in electrical discharges.

Very desirable features of a maser oscillator at infrared or optical frequencies would be a high order of monochromaticity and tunability. In the microwave range, a maser oscillator is almost inherently a very monochromatic device. However, a solid state maser can also normally be tuned over a rather large fractional variation in frequency. Both of these features are much more difficult to obtain in the infrared or optical regions. Frequencies of atomic or molecular resonances can in principle be tuned by Stark or Zeeman effects, as they would be in the radio-frequency or microwave range. However, such tuning is usually limited to a few wave numbers (or a few times 10,000 Mc/sec), which represents a large fractional change in the microwave range and only a small fractional change in the optical region. Certain optical and infrared transitions of atoms in solids are strongly affected by neighboring atoms. This may be the result of Stark effects due to internal electric fields or, as in the case of antiferromagnetic resonances, internal magnetic fields may vary enough with temperature to provide tuning over a few tens of wave numbers. Hence variation of temperature or pressure can produce some tuning. However, it appears unreasonable to expect more than a small fractional amount of tuning in an infrared or optical maser using discrete levels.

Spectrum of a Maser Oscillator

Monochromaticity of a maser oscillator is very closely connected with noise properties of the device as an amplifier. Consider first a maser cavity for optical or infrared frequencies which supports a single isolated mode. As in the microwave case, it is capable of detecting, in the limit, one or a few quanta, corresponding to a noise temperature of $h\nu/k$. However, at a wavelength of 10,000 Å, this noise temperature is about 14,000°K, and hence not remarkably low. Furthermore, other well-known photon detectors, such as a photo-electric tube, are capable of detecting a single quantum. At such frequencies, a maser has no great advantage over well-known techniques in detecting small numbers of quanta. It does offer the new possibility of coherent amplification. However, if many modes rather than a single one are present in the cavity, a rather large background of noise can occur, the noise temperature being proportional to the number of modes which are confused within the resonance width of the atomic or molecular system. A method for isolation of an individual mode which avoids this severe difficulty will be discussed below.

Let us examine now the extent to which the normal line width of the emission spectrum of an atomic system will be narrowed by maser action, or hence how monochromatic the emission from an infrared or optical maser would be. Considerations were given above concerning the number of excited systems required to produce stimulated power which would be as large as spontaneous emission due to all modes of a multimode cavity which lie within the resonance width of the system. Assume for the moment that a single mode can be isolated. Spontaneous emission into this mode adds waves of random phase to the electromagnetic oscillations, and hence produces a finite frequency width which may be obtained by analogy with expression (2) as

$$\Delta\nu_{osc} = (4\pi h\nu/P)(\Delta\nu)^2, \tag{17}$$

where $\Delta\nu$ is the half-width of the resonance at half-maximum intensity, and P the power in the oscillating field. Note that kT, the energy due to thermal agitation, has been replaced in expression (15) by $h\nu$, the energy in one quantum. Usually at these high frequencies, $h\nu \gg kT$, and there is essentially no "thermal" noise. There remains, however, "zero-point fluctuations" which produce random noise through spontaneous emission, or an effective temperature of $h\nu/k$.

For the case considered numerically above, $4\pi h\nu\Delta\nu/P$ is near 10^{-6} when P is given by expression (16), so that $\Delta\nu_{osc} \sim 10^{-\Delta}\,\Delta\nu$. This corresponds to a remarkably monochromatic emission. However, for a multimode cavity, this very monochromatic emission is superimposed on a background of spontaneous emission which has width $\Delta\nu$, and which, for the power P assumed, is of intensity equal to that of the stimulated emission. Only if the power is increased by some additional factor of about ten, or if the desired mode is separated from the large number of undesired ones, would the rather monochromatic radiation stand out clearly against the much wider frequency distribution of spontaneous emission.

Another problem of masers using multimode cavities which is perhaps not fundamental, but may involve considerable practical difficulty, is the possibility of oscillations being set up first in one mode, then in another—or perhaps of continual change of modes which would represent many sudden jumps in frequency. If the cavity dimensions, density distribution of gas and distribution of excited states remain precisely constant, it seems unlikely that oscillations will build up on more than one mode because of the usual nonlinearities which would allow the most favored mode to suppress oscillations in those which are less favored. However, if many nearby modes are present, a very small change in cavity dimensions or other characteristics may produce a shift of the oscillations from one mode to another, with a concomitant variation in frequency.

Selection of Modes for Amplification

We shall consider now methods which deviate from those which are obvious extensions of the microwave or radio-frequency techniques for obtaining maser action. The large number of modes at infrared or optical frequencies which are present in any cavity of reasonable size poses problems because of the large amount of spontaneous emission which they imply. The number of modes per frequency interval per unit volume cannot very well be reduced for a cavity with dimensions which are very large compared to a wavelength. However, radiation from these various modes can be almost completely isolated by using the directional properties of wave propagation when the wavelength is short compared with important dimensions of the region in which the wave is propagated.

Consider first a rectangular cavity of length D and with two square end walls of dimension L which are slightly transparent, its other surfaces being perfectly reflecting. Transparency of the end walls provides coupling to external space by a continuously distributed excitation which corresponds to the distribution of field strength at these walls. The resulting radiation produces a diffraction pattern which can be easily calculated at a large distance from the cavity, and which is effectively separated from the diffraction pattern due to any other mode of the cavity at essentially the same frequency.

The field distribution along the end wall, taken as the xy plane, may be proportional, for example, to $\sin(\pi rx/L)\cos(\pi sy/L)$. The resonant wavelength is of the form

$$\lambda = \frac{2}{[(q/D)^2 + (r/L)^2 + (s/L)^2]^{\frac{1}{2}}}, \tag{18}$$

where q is the number of half-wavelengths along the z direction. If L is not much smaller than D, and if $q \gg r$ or s, the resonant wavelength is approximately

$$\lambda = \frac{2D}{q}\left[1 - \frac{1}{2}\left(\frac{Dr}{Lq}\right)^2 - \frac{1}{2}\left(\frac{Ds}{Lq}\right)^2\right], \tag{19}$$

which is primarily dependent on q and insensitive to r or s. The direction of radiation from the end walls, however, is critically dependent on r and s. The Fraunhofer diffraction pattern of the radiation has an intensity variation in the x direction given by

$$I \propto (2\pi r)^2 \sin^2\left(\frac{\pi L \sin\theta}{\lambda} + \frac{\pi r}{2}\right) \Big/ \left(\pi r + \frac{2\pi L \sin\theta}{\lambda}\right)^2 \left(\pi r - \frac{2\pi L \sin\theta}{\lambda}\right)^2, \tag{20}$$

where θ is the angle between the direction of observation and the perpendicular to the end walls. For a given value of r, the strongest diffraction maxima occur at

$$\sin\theta = \pm r\lambda/2L,$$

and the first minima on either side of the maxima at

$$\sin\theta = r\lambda/2L \pm \lambda/L.$$

Thus the maximum of the radiation from a mode designated by $r + 1$ falls approximately at the half-intensity point of the diffraction pattern from the mode designated by r, which is just sufficient for significant resolution of their individual beams of radiation. This provides a method for separately coupling into or out of one or a few individual modes in the multimode cavity. A practical experimental technique for selecting one or a few modes is to focus radiation from the end walls by means of a lens onto a black screen in the focal plane. A suitable small hole in the screen will accept only radiation from the desired mode or modes.

There may, of course, be more than one mode which has similar values of r and s but different values of q, and which radiate in essentially identical directions. However, the frequencies of such modes are appreciably different, amd may be sufficiently separated from each other byan appropriate choice of the distance D between plates. Thus if only one mode with a particular value of r and s is wanted within the range of response $2\Delta\nu$ of the material used to produce oscillations, D should be less than $c/4\Delta\nu$. Or, if it is undesirable to adjust D precisely for a particular mode, and approximately one mode of this type is wanted in the range $2\Delta\nu$, one may choose

$$D \approx c/4\Delta\nu. \tag{21}$$

For the conditions assumed above, the value of D given by (21) has the very practical magnitude of about 10 cm.

It is desirable not only to be able to select radiation from a single mode, but also to make all but one or a few modes of the multimode cavity lossy in order to suppress oscillations in unwanted modes are appreciably different, and may be sufficiently separated properties. Loss may be introduced perhaps most simply by removing the perfectly reflecting walls of the cavity. The "cavity" is then reduced to partially transparent end plates and nonexistent (or lossy) perfectly matched side walls.

Suppose now that one of the modes of such a cavity is excited by suddenly introducing the appropriate field distribution on one of the end walls. This will radiate a wave into the cavity having directional properties such as those indicated by the diffraction pattern (20). If r and s have their minimum values, the maximum energy occurs near $\theta = 0$, and the wave travels more or less straight back and forth between the two plates, except for a gradual spreading due to diffraction. If r or s are larger, the maximum energy occurs at an appreciable angle θ, and the wave packet will wander off the reflecting plates and be lost, perhaps after a number of reflections. Those modes for which θ is large are highly damped and merge into a continuum, since energy radiated into them travels immediately to the walls and is lost from the cavity. However, modes for which θ is quite small may have relatively high Q and hence be essentially discrete.

For estimates of damping, consider first two end plates of infinite extent, but excited only over a square area of dimension L by a distribution which corresponds to one of our original modes. The radiated wave will be reflected back and forth many times, gradually spreading out in the diffraction pattern indicated by (20). If a mode with small values of r and s is used, the wave undergoes reflection every time it travels a distance D, and the rate of loss of energy W is given by the equations

$$dW/dt = -c(1 - \alpha)W/D,$$

$$W = W_0 e^{-c(1-\alpha)t/D} \tag{22}$$

The decay time t is then $D/c(1 - \alpha)$ rather than that given by (4) for the multimode case, or the effective distance travelled is $D/(1 - \alpha)$.

Since the wavelength for modes with small r and s is given by (19), the frequency separation between modes with successive values of q is given by the usual Fabry–Perot condition

$$\delta\nu = c/2D. \tag{23}$$

Thus $\delta\nu \gg 1/t$ and the modes with successive values of q are discrete if $1 - \alpha$, the loss on reflection, is much less than unity. On the other hand, the various modes given by small values of r or s and the same value of q are nearly degenerate, since according to (19) their frequency difference is less than $\delta\nu$ given in (23) by the factor r/q, which is of the order 10^{-4} for a typical case. These modes must be separated purely by their directional properties, rather than by their differences in frequency.

After travelling a distance $D/(1 - \alpha)$, the radiation resulting from the excitation discussed above will have moved sideways in the x direction along the infinite parallel plates a distance of approximately $D\theta/(1 - \alpha)$, where θ is the angle of one of the two large diffraction maxima given by (20). This distance is then

$$x = D\lambda r/[2(1 - \alpha)L]. \tag{24}$$

Consider now the case of finite end plates of dimension L without their infinite extension which was assumed immediately above. After a number of reflections, the diffraction pattern would no longer be precisely that given by (20). However, expression (24) would still give a reasonable approximation to the distance of sideways motion, and if this distance is larger than the end-wall dimension L, the radiation will have been lost to the cavity, and the decay time for the mode in question is appreciably shorter than that indicated by (22). This condition occurs when

$$D\lambda r/[2(1 - \alpha)L] \gtrsim L, \quad \text{or} \quad r \gtrsim 2(1 - \alpha)L^2/D\lambda. \tag{25}$$

Thus to damp out modes with $r \gtrsim 10$, when $L = \frac{1}{2}$ cm, $\alpha = 0.98$, and $\lambda = 10^{-4}$ cm, the separation D between plates needs to be as large as a few centimeters. By choosing L sufficiently small, it is possible to discriminate by such losses between the lowest mode ($r = 1$), and any higher modes. Too small a ratio $L^2/D\lambda$ will, however, begin to appreciably add to the losses from the lowest mode, and hence is undesirable if the longest possible delay times are needed.

The precise distribution of radiation intensity in the plane of the end walls which will give minimum loss, or which will occur during maser oscillation, cannot be very easily evaluated. It must, however, be somewhat like the lowest mode, $r = 1$, $s = 0$. A normal and straightforward method for exciting a Fabry–Perot interferometer is to use a plane wave moving perpendicular to the reflecting plates, and screened so that it illuminates uniformly all but the edge of the plates. Such a distribution may be expressed in terms of the nearly degenerate modes of the "cavity" with various r and s, and the considerable majority of its energy will be found in the lowest mode $r = 1$, $s = 0$, if it is polarized in the y direction. There is, of course, an exactly degenerate mode of the same type which is polarized in the x direction. Any much more complicated distribution than some approximation to uniform illumination or to the lowest mode $r = 1$, $s = 0$ of our rectangular cavity will produce a wider diffraction pattern which would be lost to a detector arranged to accept a very small angle θ near zero, and which would also be subject to greater losses when $L^2/D\lambda$ is small. However, nonuniform distribution of excited atoms between the reflecting plates could compensate for the larger diffraction losses, and in some cases induce oscillations with rather complex distributions of energy.

The above discussion in terms of modes of a rectangular cavity illustrates relations between the arrangement using a Fabry–Perot interferometer and the usual microwave resonant cavity.* An alternative approach which uses the approximation of geometrical optics more directly may also be helpful and clarifying. An atom radiating spontaneously in any direction has a decay time τ given by expression (10). The probability per unit time of emission of a quantum within a given solid angle $\Delta\Omega$ is then

$$\frac{1}{t'} = \frac{16\pi^3\nu^3}{3hc^3}\mu^2\Delta\Omega. \tag{26}$$

Hence if a sufficiently small solid angle is selected from the radiation, the amount of spontaneous emission can be made arbitrarily small. However, if essentially all the stimulated emission emitted from the end-wall of the interferometer is to be collected in a receiver or detector, allowance must be made for diffraction and a

Note added in proof.*—Use of two paralled plates for a maser operating at short wavelengths has also recently been suggested by A. M. Prokhorov (*JETP* **34, 1658 (1958)) and by R. H. Dicke (U.S. Patent 2,851,652 (September 9, 1958)). These sources do not, however, discuss the reduction of excess modes or spontaneous emission.

solid angle as large as about $(\lambda/L)^2$ must be used, so that the rate of spontaneous emission into the detector is

$$\frac{1}{t'} = \frac{16\pi^3\mu^2\nu}{3hcL^2}. \tag{27}$$

The rate of spontaneous emission (27) within the diffraction angle may be compared with the rate of induced transitions produced by one photon reflected back and forth in the volume L^2D. This rate is, as in (3), $(\mu' E/\hbar)^2(4\pi\Delta\nu)^{-1}$, where $E^2L^2D/8\pi = h\nu$. That is, since $\mu'^2 = \mu^2/3$,

$$\frac{1}{t''} = \frac{8\pi^2\mu^2\nu}{3\Delta\nu L^2 Dh}. \tag{28}$$

If D is $c/4\Delta\nu$ as in expression (21), so that there is approximately one and only one interference maximum of the interferometer in a particular direction within the range $2\Delta\nu$ of emission, then (28) becomes

$$\frac{1}{t''} = \frac{32\pi^2\mu^2\nu}{3hcL^2}. \tag{29}$$

Except for a small numerical factor of the order of the accuracy of the approximations used here, $1/t''$ given by (29) may be seen to equal $1/t'$. That is, use of the limiting amount of directional selection reduces the background of spontaneous emission to the same rate as that of stimulated emission due to a single photon. This is similar to the situation of a single mode in a cavity at microwave frequencies. It affords the limit of sensitivity which can be obtained by the usual maser amplifier, and the smallest possible noise for such a system as an oscillator.

Consider now the rate of loss of energy from a beam being reflected back and forth between the two end plates in the approximation of geometric optics. If the angle of deviation from the direction perpendicular to the plates is θ, then the additional rate of energy loss from a plane wave due to its spilling off the edges of the reflecting surfaces is

$$dW'/dt = -c\theta W/L. \tag{30}$$

Expression (30) assumes, to be precise, that the deviation is parallel to one edge of the end plate. Thus, when $\theta = (1 - \alpha)L/D$, the decay time is one-half that for $\theta = 0$. Because of nonlinearities when oscillations set in, it may be seen from expression (3) that only those modes with the largest decay times will fulfill the condition for oscillation. The fraction ε of all modes of the "cavity" which have decay times greater than one-half that of the maximum decay time is approximately $(2\theta)^2/2\pi$, or

$$\varepsilon = 2(1 - \alpha)^2 L^2/\pi D^2 . \tag{31}$$

Letting $(1 - \alpha) = 1/20$, $L = 1$ cm, and $D = 10$ cm, one obtains $\varepsilon = 1.6 \times 10^{-5}$. This enormously reduces the number of modes which are likely to produce oscillations. Since the total number of modes is, from (14), $(8\pi^2\nu^2\Delta\nu/c^3)L^2D$, this number which may produce oscillations is

$$p' = \frac{16\pi(1 - \alpha)^2\nu^2\Delta\nu L^4}{c^3 D} . \tag{32}$$

Under the assumptions used above, p' may be found to be approximately 10^5, which is very much smaller than the total number of modes in the multimode cavity, but still may be an inconveniently large number. By using limiting values of the solid angle $(2\theta)^2$ set by diffraction, the number of modes can be further reduced to approximately unity, as was seen above.

Further Discussion of Properties of Masers using Large Dimensions

It is important to notice that in the parallel plate case a very large amount of spontaneous emission may be radiating in all directions, even though only the very small amount indicated above is accepted in the detector, or is confused with the amplified wave. This property is quite different from the normal case in the microwave or radio-frequency range, and requires a rather rapid rate of supply of excited systems in order to maintain enough for maser action. Furthermore, great care must be taken to avoid scattering of light from undesired modes into the one which is desired. The fraction of spontaneous light which is scattered into the detector must be typically as low as about 10^{-6} or 10^{-7} in order to approach genuine isolation of a single mode.

Admission of a signal into the region between the two parallel plates is very similar to the process involved in a microwave cavity. The partially reflecting surfaces are analogous to coupling holes. If a monochromatic plane wave strikes the outside surface of one of the partially reflecting planes, energy will build up within the region between the planes, and the relations between input wave, energy in the "cavity", and output waves are just analogous to those for a microwave impinging on an appropriate cavity with input and output coupling holes.

Another interesting property of optical or infrared maser action which is associated with directional selection is that a beam of light may be passed through an ensemble of excited states with resulting amplification, but no important change in the wave front or phase. This amplification is just the inverse of an absorption, where it is well known that the wave front and phase are not distorted. Suppose, for example, that parallel light is focused by a lens. If an amplifying medium of excited gas is interposed between the lens and its focal point, the image will be intensified, but not otherwise changed except for some more or less normal effects which may be attributed to the dielectric constant of the gas. The same situation can in principle, occur for maser amplification of microwaves. However, at these lower frequencies the amplification per unit length is usually so small that an impractically large volume of excited material would be required for amplification of a wave in free space to be evident. There may be a considerable amount of spontaneous emission in all directions, but only a very small fraction of the total spontaneous emission will fall at the focal point of the lens and be superimposed as noise on the intensified image. Noise from spontaneous emission decreases, for example, with the inverse square of distance from the emitting material, whereas the intensity of the focused beam increases as one approaches the focal point.

A Specific Example

As an example of a particular system for an infrared maser, let us consider potassium. Atomic potassium is easily vaporized and has a simple spectrum as indicated by the energy levels shown in Fig. 1. Absorption transitions can occur from the $4s\ ^2S_{\frac{1}{2}}$ ground state only to the various p levels. In particular, the atoms can be excited to the $5p\ ^2P_{\frac{1}{2},\frac{3}{2}}$ by radiation of wavelength 4047 Å. Just the right exciting frequency can be obtained from another potassium lamp, whose light is filtered to remove the red radiation at 7700 Å. These excited atoms will decay to the $5s$ or $3d$ states in about 2×10^{-7} sec, or more slowly to the $4s$ ground state. However, if excited atoms are supplied fast enough, a sizable population can be maintained in the $5p$ state.

The minimum number of excited atoms required for maser-type oscillation may be found from (6), if the dipole matrix element were known, or from (11) if the lifetime were known. Although the wave functions necessary for obtaining the matrix elements have been calculated,[5,6] only estimates of the matrix element or lifetime can be made at present. The rate at which atoms must be supplied may, however, be obtained without detailed knowledge of the matrix elements by a small modification of expression (11). If τ is the mean life for spontaneous radiation of the desired wavelength, and ϕ is the fraction of the all decay processes from the upper level which occur in this manner, then the actual mean life in the excited state is $\tau' = \phi\tau$. The number of atoms needed per second can be obtained from (11) as

$$\frac{n}{\tau'} = \frac{p}{\phi t} = \frac{8\pi^2 V}{(\pi \ln 2)^{\frac{1}{2}}\phi t \lambda^3}\frac{\Delta\nu}{\nu} = 8\pi(2\pi)^{\frac{1}{2}}\frac{V}{\phi\lambda^3 ct}\left(\frac{kT}{m}\right)^{\frac{1}{2}}, \qquad (33)$$

where λ is the wavelength. Thus, if the fraction ϕ is known, no other detailed properties of the atomic transition are required to evaluate the rate at which excited atoms must be supplied.

[5] L. Biermann and K. Lubeck, *Z. Astrophys.* **25**, 325 (1948).
[6] D. S. Villars, *J. Opt. Soc. Am.* **42**, 552 (1952).

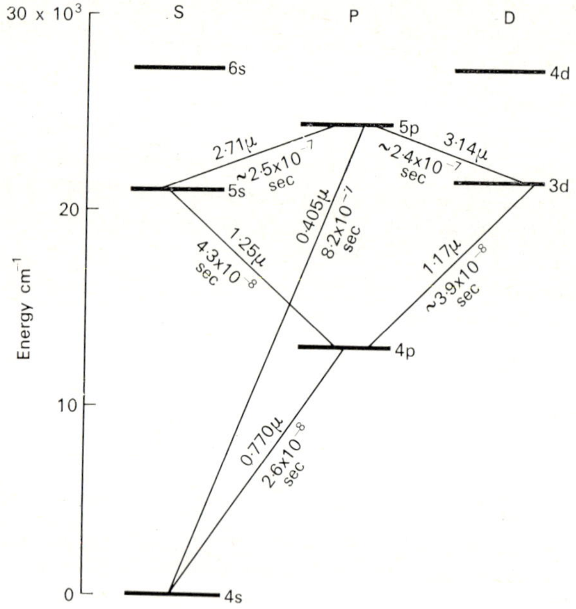

FIG. 1. Low-lying energy levels and transitions of atomic potassium.

For the particular case of a gas (such as potassium vapor) at sufficiently low pressure that collisions are not too frequent, we can obtain ϕ from the relative intensities of the various radiative transitions out of the excited state. Observed relative intensities[7] show that for potassium the $5p \rightarrow 3d$ transitions are about 4 times more intense than the $5p \rightarrow 5s$ transitions. Within the $5p \rightarrow 3d$ transitions it follows from elementary angular momentum theory that the $5p_{\frac{3}{2}} \rightarrow 3d_{\frac{5}{2}}$ is the most intense, accounting for 9/15 of the radiation emitted. Using the observed intensity ratio to allow for transitions to the 5s level we conclude that about $9/18 = \frac{1}{2}$ of those atoms excited to the $5p_{\frac{3}{2}}$ level decay to the $3d_{\frac{5}{2}}$ level. Decay to

[7] *Viz.*, Tabulation in the *Handbook of Chemistry and Physics*, edited by D. Hodgman (Chemical Rubber Publishing Company, Cleveland, 1957), thirty-ninth edition.

the 4*s* ground state is almost certainly less likely since we do know that this matrix element is not very large ($f = 0.010$,[5] so that $\mu = 0.65 \times 10^{-18}$ esu). Thus $\phi = \frac{1}{2}$ for the transition $5p_{\frac{3}{2}} - 3d_{\frac{5}{2}}$ at 31, 391 Å.

Assume now two parallel plates of area 1 cm² and 10 cm apart, having a reflectivity α of 0.98. The decay time t for radiation in the space between the plates is $[10/(3 \times 10^{10})] \times 50$ sec and $V = 10\ \text{cm}^3$. For potassium vapor of suitable pressure, $T = 435°\text{K}$ and $m = 39$ amu. Hence, from (29), the number of excited atoms needed per second is $dn/dt \geq 2.5 \times 10^{15}$.

The energy needed per second is $d/dt(nh\nu)$, where ν is the frequency of the exciting radiation. Its value is 1.2×10^{-3} watt. This energy requirement is quite attainable. Incomplete absorption of the exciting radiation, reflection losses and multiplicity of the atomic states might raise this requirement somewhat. The absorption of the existing radiation is easily calculable and can be adjusted by controlling the density of the vapor:

$$k_0 = \frac{1}{\Delta\nu_D}\left(\frac{\ln 2}{\pi}\right)^{\frac{1}{2}} \frac{\pi e^2}{mc} Nf, \tag{34}$$

where k_0 is the absorption coefficient at the peak of the line $\Delta\nu_D$ is the (Doppler) line half-width, e is the electron charge, m is the electron mass, c is the velocity of light, N is the number of initial state atoms per cc, and f is the oscillator strength of the transition; i.e.,

$$k_0 = 1.25 \times 10^{-2}(Nf/\Delta\nu_D).$$

For the exciting transition, 4046 Å, in potassium, $\nu_0 = 7.42 \times 10^{14}$ cycles/sec and at 435°K, $\Delta\nu_D = 0.84 \times 10^9$ cycles/sec. At this temperature the vapor pressure is 10^{-3} mm of mercury, so that in saturated vapor $N = 2.5 \times 10^{13}$/cc. Since $f = 0.10$ for the $4s_{\frac{1}{2}} \rightarrow 5p_{\frac{3}{2}}$ transition,[6] $k_0 = 3.72$. This is high enough that the exciting radiation would be absorbed in a thin layer; if necessary it can be reduced by changing the pressure or temperature.

The light power for excitation is proportional to

$$\frac{V}{t} = \frac{AD}{(1 - \alpha)D/c} = \frac{Ac}{1 - \alpha}, \tag{35}$$

where V is the volume of the cavity, t is the decay time for light in the cavity, D is the length of the cavity, A is the cross-section area of the cavity, α is the reflectivity of the end plates, and c is the velocity of light. This is independent of length, so that for a given cross-sectional area the light density needed can be reduced by increasing the length.

Light Sources for Excitation

A small commercial potassium lamp (Osram) was operated with an input of 15 watts, 60 cycles, and its output was measured. In the red lines (7664–7699 Å), the total light output was 28 mw from about 5 cc volume. At the same time, the total output in the violet lines (4044–4047 Å) was 0.12 mw,[8] so that the output in $4s - 5p_{\frac{3}{2}}$ was 0.08 mw. By increasing the current from 1.5 to 6 amp, with forced air cooling (the outer jacket being removed), the total violet output was increased to 0.6 mw. These outputs are somewhat short of the power level needed, but they may be considerably increased by adjusting discharge conditions to favor production of the violet line, and by using microwave excitation. With a long maser cell, the lamp area can be greatly increased. If necessary, very high peak light powers could be obtained in pulsed operation, although one would have to be careful not to broaden the line excessively.

Another possibility for excitation is to find an accidental coincidence with a strong line of some other element. The $8p$ level of cesium is an example of this type, since it can be excited very well by a helium line. The 4047 Å line of mercury is 5 cm^{-1} from the potassium line, and is probably too far away to be useful even when pressure broadened and shifted.

[8] We are indebted to R. J. Collins for making these measurements.

Different modes correspond to different directions of propagation, and we only want to produce one or a few modes. Thus the cavity need only have two good reflecting walls opposite each other. The side walls need not reflect at all, nor do they need to transmit infrared radiation.

Unfortunately, most elements which have simple spectra, are quite reactive. Sapphire has good chemical inertness and excellent infrared transmission, being almost completely transparent as far as about 4 microns wavelength.[9] With such good transmission, the principal reflecting surfaces can be put outside the cell, and hence chosen for good reflectivity without regard to chemical inertness. Thus, one could use gold which has less than 2% absorption in this region, and attain a reflectivity of ~97% with 1% transmission. Even better reflectivity might be obtained with multiple dielectric layers of alternately high- and low-dielectric constant. The inner walls of the sapphire cell would reflect about 5 % of the infrared light, and the thickness should be chosen so that the reflections from the two surfaces are in phase. The phase angle between reflections from the two surfaces depends on the thickness and the refractive index. Since sapphire is crystalline and the index is different for ordinary and extraordinary rays, the thickness could be chosen to give constructive interference for one polarization, and destructive interference for the perpendicular polarization. Thus, one could discriminate, if desired, between modes traveling in the same direction with different polarization.

To select just one from among the very many modes possible within the line width, the stimulated emission of radiation with one chosen direction of propagation must be favored. Thus the cell should be made long in the desired direction and fitted with highly reflecting end plates. The desired wave then has a long path as it travels back and forth, and so has a good chance to pick up energy from the excited atoms. A large width decreases the angular discrimination, and increases the pumping power needed.

[9] R. W. Kebler, *Optical Properties of Synthetic Sapphire* (Linde Company, New York).

For the potassium radiation at 3.14×10^{-4} cm wavelength, and $\Delta\nu$ being the Doppler width at 435°K, i.e., $\Delta\nu/\nu_0 = 1.2 \times 10^{-6}$, the number of modes is $2.0 \times 10^6\ V$ from expression (15). If we consider a cavity 1 cm square by 10 cm long, this number is 2.0×10^7, or 3.2×10^6 modes per steradian (forward and backward directions are taken as equivalent for standing waves). The angular separation between modes is then $(32 \times 10^6/2)^{-\frac{1}{2}} = 2.5 \times 10^{-4}$ radian, where the 2 in the denominator removes the polarization degeneracy. The angular aperture accepted by this cavity is 1/10, but if the end plates had 98% reflectivity, the effective length would be increased by a factor of 50, and the angular aperture reduced to 2×10^{-3} radian. Thus there would be only 8 modes of each polarization within the effective aperture of the cell. Obviously this type of mode selection could be pushed further by making the cavity longer or narrower or more reflecting but this should not be necessary. Furthermore, the emission line does not have constant intensity over the width $\Delta\nu$, and the mode nearest the center frequency would be the first to oscillate at the threshold of emission.

Solid-state Devices

There are a good many crystals, notably rare earth salts, which have spectra with sharp absorption lines, some of them having appeared also in fluorescence. In a solid, a concentration of atoms as large as 10^{19} per cc may be obtained. The oscillator strengths of the sharp lines are characteristically low, perhaps 10^{-6}. If the f value is low, radiative lifetimes are long, and in some cases lifetimes are as long as 10^{-3} sec or even more.

If the lifetime is primarily governed by radiation in the desired line, the pumping power required for the onset of stimulated oscillation is independent of the f value, as was shown above. For the atomic potassium level considered earlier, there are several alternative radiative decay paths (to the $4s$ and $3d$ states). In a solid there may also be rapid decay by nonradiative procesess. If the storage time is long, because of a small f value, there is more time for competing processes to occur. Even lines which are sharp for solids are likely to be broader than those obtainable in gases. This larger width makes the attainment of maser oscillation more difficult, and it adds greatly to the difficulty of selecting a single mode. However, there may very well be suitable transitions among the very many compounds.

The problem of populating the upper state does not have as obvious a solution in the solid case as in the gas. Lamps do not exist which give just the right radiation for pumping. However, there may be even more elegant solutions. Thus it may be feasible to pump to a state above one which is metastable. Atoms will then decay to the metastable state (possibly by nonradiative processes involving the crystal lattice) and accumulate until there are enough for maser action. This kind of accumulation is most likely to occur when there is a substantial empty gap below the excited level.

Summary and High-frequency Limits

The prospect is favorable for masers which produce oscillations in the infrared or optical regions. However, operation of this type of device at frequencies which are still very much higher seems difficult. It does not appear practical to surround an atomic system with cavity walls which would very much affect its rate of spontaneous emission at very short wavelengths. Hence any ensemble of excited systems which is capable of producing coherent amplification at very high frequencies must also be expected to emit the usual amount of spontaneous emission. The power in this spontaneous emission, from expressions (14) and (16), increases very rapidly with frequency—as ν^4 if the width $\Delta\nu$ is due to Doppler effects, or as ν^6 if the width is produced by spontaneous emission. By choice of small matrix elements, $\Delta\nu$ can, in principle, be limited to that associated with Doppler effects, but the increase in spontaneously emitted power as fast as ν^4 is unavoidable.

For a wavelength $\lambda = 10^4$ Å, it was seen above that spontaneous emission produced a few milliwatts of power in a maser system of dimensions near one centimeter, assuming reflectivities which seem attainable at this wavelength. Thus in the ultraviolet region at $\lambda = 1000$ Å, one may expect spontaneous emissions of intensities near ten watts. This is so large that supply of this much power by excitation in some other spectral line becomes very difficult. Another decrease of a factor of 10 in λ would bring the spontaneous emission to the clearly prohibitive value of 100 kilowatts. These figures show that maser systems can be expected to operate successfully in the infrared, optical, and perhaps in the ultraviolet regions, but that, unless some radically new approach is found, they cannot be pushed to wavelengths much shorter than those in the ultraviolet region.

For reasonably favorable maser design in the short wavelength regions, highly reflecting surfaces and means of efficient focusing of radiation must be used. If good reflecting surfaces are not available, the number of excited systems used must, from (6) be very much increased with a resulting increase in spontaneous emission and difficulty in supply of excited systems. If focusing is not possible, the directional selection of radiation can in principle be achieved by detection at a sufficiently large distance from the parallel plates. However, without focusing, the directional selection is much more difficult, and the background of spontaneous emission may give serious interference as noise superimposed on the desired readiation.

Finally, it must be emphasized that, as masers are pushed to higher frequencies, the fractional range of tunability must be expected to decrease more or less inversely as the frequency. The absolute range of variation can be at least as large as the width of an individual spectral line, or as the few wave numbers shift which can be obtained by Zeeman effects. However, continuous tuning over larger ranges of frequency will require materials with very special properties.

Acknowledgements

The authors wish to thank W. S. Boyle, M. Peter, A. M. Clogston, and R. J. Collins for several stimulating discussions.

2

Molecular Amplifier and Generator for Submillimeter Waves*

A. M. PROKHOROV

P. N. Lebedev Physics Institute, Academy of Sciences, U.S.S.R.

IN THE present paper we consider the possibility of constructing a molecular amplifier and generator (MAG), for waves shorter than 1 mm, using ammonia molecules, The rotational transitions of the NH_3 molecules lie in the wavelength region below 1 mm. These transitions can be used to construct the MAG. The rotational transitions are sorted out at the same time as the inversion levels, viz., molecules in the lower inversion level are sorted out by passing the molecular beam through a quadrupole condenser.

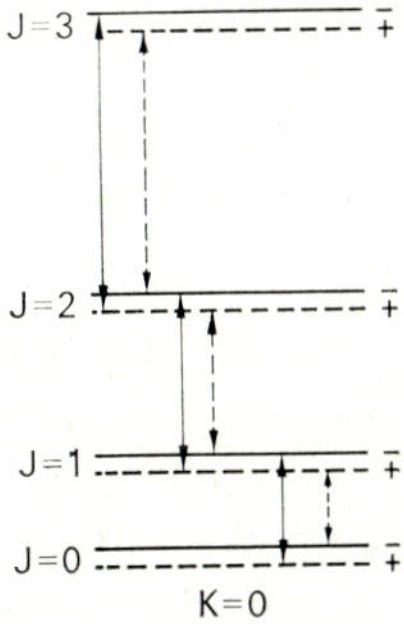

FIG. 1. The rotational spectrum of NH_3 for $J = 3, 2, 1$, and 0, and for $K = 0$.

J. Exptl. Theoret. Phys. (U.S.S.R.)* **34, 1658 (1958); *Soviet Phys. — JETP* **7,** 1140 (1958).

The system of rotational-inversion levels after sorting is given in the figure for $J = 3, 2, 1$, and 0 and for $K = 0$. Levels which are not occupied by molecules are shown by dotted lines. The solid arrows show transitions increasing the energy of the incident radiation; dotted arrows show those absorbing energy.

An amplifier can be constructed using a device in which the radiation coming from one horn crosses a number of molecular beams and falls on a second horn. If the average density of the number of active molecules is equal to N, the coefficient of negative absorption is determined by the equation

$$\alpha = 8\pi^2 \nu \, |\mu_{mn}|^2 \, N/hc\Delta\nu \,, \tag{1}$$

where ν is the frequency of the transition, μ_{mn} the dipole-moment matrix element, $\Delta\nu$ the line width, h Planck's constant, and c the velocity of light.

If the power of the radiation leaving horn 1 is equal to P_0, the power after passing a path l and entering horn 2 rises to $P_k = P_0 e^{\alpha l}$. Let $\nu = 6 \times 10^{11}$ cps ($\lambda = 0.5$ mm), $|\mu_{mn}|^2 = 2 \times 10^{-36}$, $\Delta\nu = 5 \times 10^3$ cps, and $N = 10^{10}$ cm^{-3}. Then $\alpha = 1$ cm^{-1}. If $l = 10$ cm, $P_k/P_0 + 2.2 \times 10^4$. The maximum power which such a beam can produce is about one microwatt. To construct a molecular generator one can use two plane-parallel mirrors as the resonator. If the distance between the mirrors is l, the reflection coefficient of the mirrors is k, and we assume that energy losses of the place waves occur only upon reflection from the mirrors, the Q-factor of such a system is equal to

$$Q = (2\pi l/\lambda)/(1 - k) \,. \tag{2}$$

If $l = 1$ cm, $\lambda = 0.05$ cm, $k = 0.95$, then $Q = 2400$. However, energy losses occur also because the wave is not plane but has an angular spread $2\theta \approx \lambda/D$, where D is the linear dimension of the mirror. Because of this effect, the energy P_n after n reflections will be

$$P_n = P_0/(1 + nl\lambda/D^2)^2 \,. \tag{3}$$

The quantity nl is the path traversed by the wave during the n reflections. This time is equal to $\tau = nl/c$.

If we know the Q-factor of the system, the time in which the power decreases by a factor e is equal to $\tau = Q/2\pi\nu$. During this time the wave traverses a path $nl = c\tau$; if $Q = 2400$, and $\nu = 6 \times 10^{11}$ cps, $nl = 21$ cm.

If $D = 3$ cm, we get from (3) $P_n = 0.8\, P_0$, i.e., in our case the losses during reflection play the dominant part.

The condition for self-excitation can be written in the form

$$ke^{\alpha l} > 1. \tag{4}$$

If $\alpha = 1\ \text{cm}^{-1}$, $l = 1$ cm, $k = 0.95$, condition (4) is satisfied by a wide margin. If $e^{\alpha l} \gg 1$, self-excitation occurs for small k.

3

Resonant Modes in a Maser Interferometer*

A. G. Fox and Tingye Li

Summary

A theoretical investigation has been undertaken to study diffraction of electromagnetic waves in Fabry–Perot interferometers when they are used as resonators in optical masers. An electronic digital computer was programmed to compute the electromagnetic field across the mirrors of the interferometer where an initially launched wave is reflected back and forth between the mirrors.

It was found that after many reflections a state is reached in which the relative field distribution does not vary from transit to transit and the amplitude of the field decays at an exponential rate. This steady-state field distribution is regarded as a normal mode of the interferometer. Many such normal modes are possible depending upon the initial wave distribution. The lowest-order mode, which has the lowest diffraction loss, has a high intensity at the middle of the mirror and rather low intensities at the edges. Therefore, the diffraction loss is much lower than would be predicted for a uniform plane wave. Curves for field distribution and diffraction loss are given for different mirror geometries and different modes.

Since each mode has a characteristic loss and phase shift per transit, a uniform plane wave which can be resolved into many modes cannot, properly speaking, be resonated in an interferometer. In the usual optical interferometers, the resolution is too poor to resolve the individual mode resonances and the uniform plane wave distribution may be maintained approximately. However, in an oscillating maser, the lowest-order mode should dominate if the mirror spacing is correct for resonance.

A confocal spherical system has also been investigated and the losses are shown to be orders of magnitude less than for plane mirrors.

I. Introduction

Schawlow and Townes[1] have proposed infrared and optical masers using Fabry–Perot interferometers as resonators. Very

*Bell System Tech. J. **40**, 453 (1961).

recently, Maiman[2] and Collins *et al.*[3] have demonstrated experimentally the feasibility of stimulated optical radiation in ruby. In these experiments two parallel faces of the ruby sample were polished and silvered so as to form an interferometer. The radiation due to stimulated emission resonates in the interferometer and emerges from a partially silvered face as a coherent beam of light.

In a maser using an interferometer for a resonator, a wave leaving one mirror and traveling toward the other will be amplified as it travels through the active medium. At the same time it will lose some power due to scattering by inhomogeneities in the medium. When the wave arrives at the second mirror some power will be lost in reflection due to the finite conductivity of the mirror and some power will be lost by radiation around the edges of the mirror. For oscillation to occur, the total loss in power due to density scattering, diffractive spillover and reflection loss must be less than the power gained by travel through the active medium. Thus diffraction loss is expected to be an important factor, both in determining the start-oscillation condition, and in determining the distribution of energy in the interferometer during oscillation.

While it is common practice to regard a Fabry–Perot interferometer as being simultaneously resonant for uniform plane waves traveling parallel to the axis and at certain discrete angles from the axis, this picture is not adequate for the computation of diffraction loss in a maser. It is true that, when the interferometer is operated as a passive instrument with uniform plane waves continuously supplied from an external source, the internal fields may be essentially those of uniform plane waves. In an oscillating maser where power is supplied only from within the interferometer, the recurring loss of power from the edges of a wave due to diffraction causes a marked departure from uniform amplitude and phase across the mirror.

The purpose of our study is to investigate the effects of diffraction on the electromagnetic field in a Fabry–Perot interferometer in free space. The conclusions can be applied equally well to gaseous or solid state masers provided the interferometer is immersed in the active medium, i.e., there are no side-wall discontinuities.

II. Formulation of the Problem

2.1 *General Information*

Our approach is to consider a propagating wave which is reflected back and forth by two parallel plane mirrors, as shown in Fig. 1(a). [This is equivalent to the case of a transmission medium comprising a series of collinear identical apertures cut into parallel and equally spaced black (perfectly absorbing) partitions of infinite extent, as in Fig. 1(b).] We assume at first an arbitrary initial field distribution at the first mirror and proceed to compute the field produced at the second mirror as a result of the first transit. The newly calculated field distribution is then used to compute the field produced at the first mirror as a result of the second transit. This computation is repeated over and over again for subsequent transits. The questions we have in mind are: (a) whether, after many transits, the relative field distribution approaches a

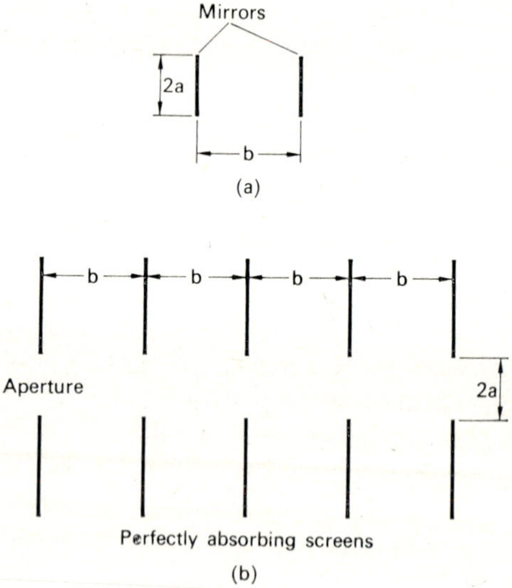

FIG. 1. The Fabry–Perot interferometer and the transmission medium analog.

steady state; (b) whether, if a steady-state distribution results, there are any other steady-state solutions; and (c) what the losses associated with these solutions would be. While it is by no means obvious that steady-state solutions (corresponding to normal modes) exist for a system which has no side-wall boundaries, it will be shown that such solutions do indeed exist.*

We shall use the scalar formulation of Huygens' principle to compute the electromagnetic field at one of the mirrors in terms of an integral of the field at the other. This is permissible if the dimensions of the mirror are large in terms of wavelength and if the field is very nearly transverse electromagnetic and is uniformly polarized in one direction. Later, we shall show that these assumptions are consistent with the results of our solutions and therefore are justifiable. We shall also show that other polarization configurations can be constructed from the solutions of the scalar problem by linear superposition.

The Fresnel field u_p due to an illuminated aperture A is given by the surface integral[4]

$$u_p = \frac{jk}{4\pi} \int_A u_a \frac{e^{-jkR}}{R} (1 + \cos\theta)\, dS, \tag{1}$$

where u_a is the aperture field, k is the propagation constant of the medium, R is the distance from a point on the aperture to the point of observation and θ is the angle which R makes with the unit normal to the aperture. We now assume that an initial wave of distribution u_p is launched at one of the mirrors of the interferometer and is allowed to be reflected back and forth in the interferometer, After q transits the field at a mirror due to the reflected field at the other is simply given by (1) with u_p replaced by u_{q+1}, which is the field across the mirror under consideration and u_a by u_q, which is the reflected field across the opposite mirror giving rise to u_{q+1}.

*Schawlow and Townes[1] suggested the possibility that resonant modes for a parallel plate interferometer might be similar in form to those for a totally enclosed cavity.

It is conceivable that after many transits the distribution of field at the mirrors will undergo negligible change from reflection to reflection and will eventually settle down to a steady state. At this point the fields across the mirrors become identical except for a complex constant; that is,

$$u_q = \left(\frac{1}{\gamma}\right)^q v, \tag{2}$$

where v is a distribution function which does not vary from reflection to reflection and γ is a complex constant independent of position coordinates. Substituting (2) in (1) we have the integral equation

$$v = \gamma \int_A Kv \, dS \tag{3}$$

in which the kernel of the integral equation, K, is equal to $(jk/4\pi R) \cdot (1 + \cos\theta)e^{-jkR}$. The distribution function v, which satisfies (3) can be regarded as a normal mode of the interferometer defined at the mirror surface, and the logarithm of γ, which specifies the attenuation and the phase shift the wave suffers during each transit, can be regarded as the propagation constant associated with the normal mode.

The integral equation (3) can be solved numerically by the method of successive approximations (Ref. 5, p. 421). It is interesting to note that this iterative method of solution is analogous to the physical process of launching an initial distribution of wavefront in the interferometer and letting it bounce back and forth between the mirrors as described in the foregoing paragraphs.

We have studied and obtained numerical solutions for several geometric configurations of the interferometer. These are (a) rectangular plane mirrors, (b) circular plane mirrors and (c) confocal spherical or paraboloidal mirrors.

2.2 *Rectangular Plane Mirrors*

When the mirror separation is very much larger than the mirror dimensions the problem of the rectangular mirrors reduces to a two-dimensional problem of infinite strip mirrors. This is shown in Appendix A. The integral equation for the problem of infinite strip mirrors, when $a^2/b\lambda$ is much less than $(b/a)^2$, is

$$v(x_2) = \gamma \int_{-a}^{a} K(x_2, x_1) v(x_1)\, dx_1 \tag{4}$$

with

$$K(x_2, x_1) = \frac{e^{j(\pi/4)}}{\sqrt{\lambda b}} e^{-jk(x_1 - x_2)^2/2b} \tag{4a}$$

The various symbols are defined in Fig. 2 and Appendix A.

Equation (4) is a homogeneous linear integral equation of the second kind. Since the kernel is continuous and symmetric $[K(x_2, x_1) = K(x_1, x_2),]$ its eigenfunctions v_n corresponding to

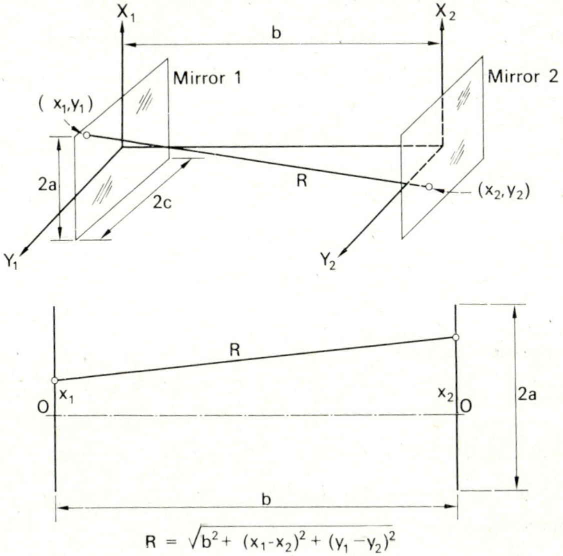

FIG. 2. Geometry of rectangular plane mirrors.

distinct eigenvalues γ_n are orthogonal in the interval $(-a,a)$; that is (Ref. 5, p. 413),

$$\int_{-a}^{a} v_m(x)v_n(x)\,dx = 0\,, \qquad m \neq n\,. \tag{5}$$

It should be noted that the eigenfunctions are in general complex and are defined over the surface of the mirrors only. They are not orthogonal in the power (Hermitian) sense as commonly encountered in lossless systems. Here, the system is basically a lossy one and the orthogonality relation is one which is generally applicable to lossy systems, such as lossy-wall waveguides.

The eigenfunctions are distribution functions of the field over mirror surfaces and represent the various normal modes of the system. The normal modes for rectangular plane mirrors are obtained by taking the products of the normal modes for infinite strip mirrors in x and y directions; that is,

$$v_{mn}(x,y) = v_{x,m}(x)v_{y,n}(y)\,. \tag{6}$$

We designate this as the TEM_{mn} mode for the rectangular plane-mirror interferometer. In view of (5) we see that the normal mode distribution functions v_{mn} are orthogonal over the surface of the rectangular mirror.

The logarithms of the eigenvalues represent propagation constants associated with the normal modes. The propagation constant for the TEM_{mn} mode of rectangular plane mirrors is given by

$$\log \gamma_{mn} = \log \gamma_{x,m} + \log \gamma_{y,n}\,. \tag{7}$$

The real part of the propagation constant specifies the loss per transit and the imaginary part the phase shift per transit, in addition to the geometrical phase shift, for the normal modes.

2.3 *Circular Plane Mirrors*

It is shown in Appendix B that the solutions to the integral equation for circular plane mirrors (Fig. 3) when $a^2/b\lambda$ is much less than $(b/a)^2$, are given by

$$v(r,\phi) = R_n(r)e^{-jn\phi} \qquad (n = \text{integer})\,, \tag{8}$$

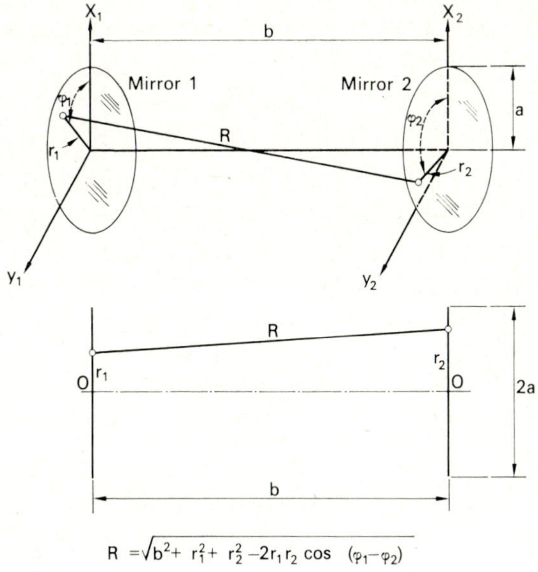

FIG. 3. Geometry of circular plane mirrors.

where $R_n(r)$ satisfies the reduced integral equation

$$R_n(r_2)\sqrt{r_2} = \gamma_n \int_0^a K_n(r_2, r_1) R_n(r_1)\sqrt{r_1},dr_1\,, \tag{9}$$

with

$$K_n(r_2, r_1) = j^{n+1}\frac{k}{b} J_n\left(k\frac{r_1 r_2}{b}\right)\sqrt{r_1 r_2}\, e^{-jk(r_1{}^2+r_2{}^2)/2b}, \tag{9a}$$

where J_n is a Bessel function of the first kind and nth order. As in the problem of infinite strip mirrors, (9) is a homogeneous linear integral equation of the second kind with a continuous and symmetric kernel. Its eigenfunctions corresponding to distinct eigenvalues are orthogonal in the interval $(0,a)$; that is,

$$\int_0^a R_{nl}(r)R_{nm}(r)r\,dr = 0, \qquad (l \neq m). \tag{10}$$

Therefore, we see that the distribution functions $v_{nm}(r,\phi) = R_{nm}(r)e^{-jn\phi}$ corresponding to distinct eigenvalues γ_{nm} are orthogonal over the surface of the mirror; that is,

$$\int_0^{2\pi}\int_0^a v_{nm}(r,\phi)v_{kl}(r,\phi)r\,dr\,d\phi = 0 \qquad (\text{either } n \neq k \text{ or } m \neq l). \tag{11}$$

The set of eigenfunctions R_{nm} describes the radial variations of field intensity on the circular mirrors, and the angular variations are sinusoidal in form. We designate a normal mode of the circular plane mirrors as the TEM_{nm} mode, with n denoting the order of angular variation and m denoting the order of radial variation. The propagation constant associated with the TEM_{nm} mode is simply $\log \gamma_{nm}$, which must be obtained from the solution of (9).

2.4 *Confocal Spherical or Paraboloidal Mirrors*

A number of geometries other than plane parallel mirrors have been suggested, and it is believed that most of these can be studied using the same iterative technique. One of the geometries we investigated is that of a confocal spherical system.[6] In this geometry the spherical mirrors have identical curvatures and their foci are coincident, as shown in Fig. 4. One of the possible advantages of such a system is the relative ease of adjustment, since the mirrors are no longer required to be parallel as in the case of the parallel plane system. Another is that the focusing action of the mirrors might give rise to lower diffraction losses.

A spherical mirror with a small curvature approximates closely a paraboloidal mirror. In the case of confocal spherical mirrors, the conditions that its curvature be small is equivalent to saying that the separation between mirrors is large compared to the dimensions of the mirrors. It is shown in Appendix C that the solutions to the integral equation for confocal paraboloidal mirrors, when $a^2/b\lambda$ is much less than $(b/a)^2$, are given by

$$v(r,\phi) = S_n(r)e^{-jn\phi} \qquad (n = \text{integer}), \tag{12}$$

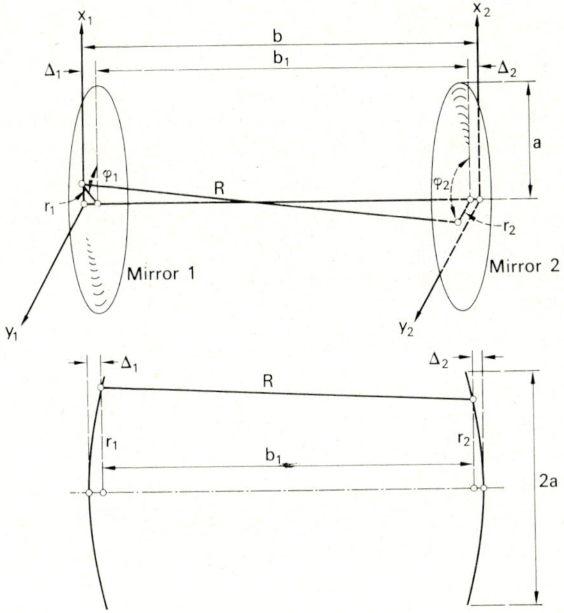

$$R = \sqrt{b_1^2 + r_1^2 + r_2^2 - 2r_1 r_2 \cos\ (\varphi_1 - \varphi_2)}$$
$$\Delta_1 = b - \sqrt{b^2 - r_1^2}$$
$$\Delta_2 = b - \sqrt{b^2 - r_2^2}$$

FIG. 4. Geometry of confocal spherical mirrors.

where $S_n(r)$ satisfies the reduced integral equation

$$S_n(r_2)\sqrt{r_2} = \gamma_n \int_0^a K_n(r_2, r_1) S_n(r_1)\sqrt{r_1}\, dr_1 , \tag{13}$$

with

$$K_n(r_2, r_1) = j^{n+1} \frac{k}{b} J_n\left(k \frac{r_1 r_2}{b}\right)\sqrt{r_1 r_2}. \tag{13a}$$

Again, we see that (13) is a homogeneous linear integral equation of the second kind with a continuous and symmetric kernel. Therefore, general remarks concerning the normal modes of circular plane mirrors given in the foregoing section are also applicable to confocal spherical or paraboloidal mirrors.

III. Computer Solutions

3.1 *General*

An IBM 704 computer was programmed to solve the integral equations for the various geometries of the interferometer by the method of successive approximations. As mentioned previously, this is analogous to the physical process of launching an initial distribution of wavefront in the interferometer and letting it bounce to and fro between the mirrors.

3.2 *Infinite Strip Mirrors*

The first problem put on the computer was that of a pair of infinite strip mirrors, having the dimensions $2a = 50\lambda$, $b = 100\lambda$. Equation (26) was employed for the computation, using an initial excitation of a uniform plane wave at the first mirror. A total of one hundred increments was used for the numerical integration. After the first transit the field intensity (electric or magnetic) had the amplitude and phase shown in Fig. 5. In these and subsequent amplitude and phase distributions the curves are normalized so that the maximum amplitude is unity, and the phase at that point is zero. The large ripples are due to the fact that the initial wave front contains 6.25 Fresnel zones as seen from the center of the second mirror. Therefore, in passing from the center to the edge of the second mirror there is a change of 3×6.25 Fresnel zones, and this agrees with the number of reversals in curvature seen in the amplitude distribution.

With subsequent transits, these ripples grow smaller, the amplitude at the edge of the mirror decreases, and the relative field distributions approach a steady state. By the time the wave had made three hundred bounces, the fluctuations occurring from bounce to bounce were less than 0.03 per cent of the final average value. The amplitude and phase for the 300th bounce are also shown in Fig. 5.

We regard this field distribution as an iterative normal mode of the interferometer. In other words, if this distribution is

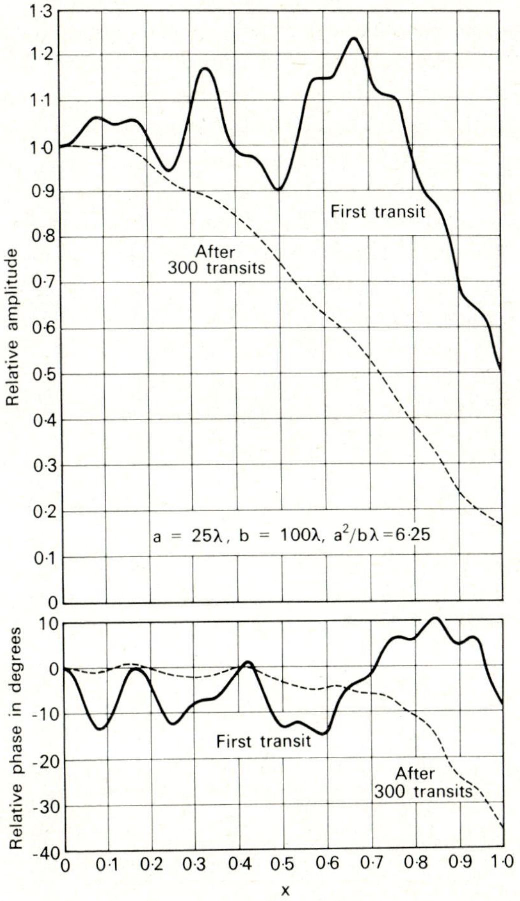

FIG. 5. Relative amplitude and phase distributions of field intensity for infinite strip mirrors. (The initially launched wave has a uniform distribution.)

introduced as an initial wave at one mirror it will reproduce the same distribution at the other mirror. Indeed, this is what the computer is verifying when we compute the 301st bounce.

Once the solutions have reached a steady-state, we can pick any point on the wavefront, say the center of the mirror, and examine how the absolute phase and amplitude change from bounce to bounce. In this way we determined that the power loss of this mode is 0.688 per cent per transit and the phase shift per transit has a lead of 1.59 degrees. Since phase shift is measured relative to the free-space electrical length between the mirrors (360 b/λ degrees), this means that the mode has an effective phase velocity which is slightly greater than the speed of light, just as for a metal tube waveguide.

In Fig. 6 is shown how the field intensity at an arbitrary off-center point ($x = 0.5a$) approaches its steady-state normalized value after a start from a uniform plane wave. After the 100th

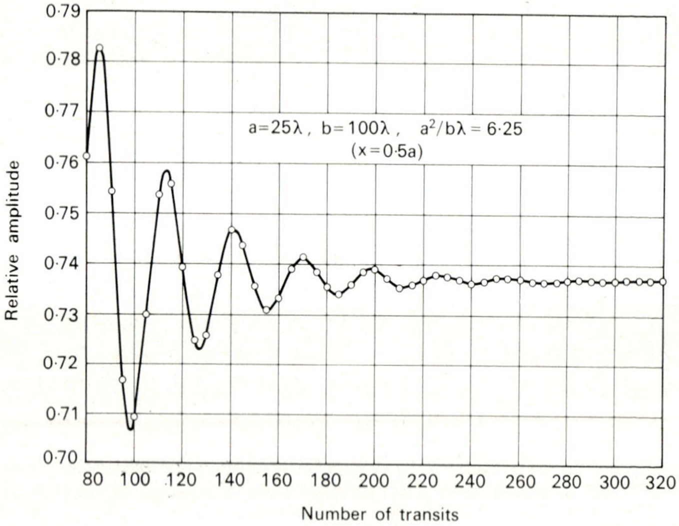

FIG. 6. Fluctuation of field amplitude at $x = 0.5$ as a function of number of transits. (The initially launched wave has a uniform distribution.)

transit the plot appears to be a damped sine wave. We interpret this damped oscillation as the beating between two normal modes having different phase velocities. The mode with the lower attenuation, of course, survives the longest, and this is the one shown in Fig. 5. We regard this as the dominant mode of the interferometer. We believe the other mode which beats with the dominant mode to be the next-higher order, even-symmetric mode. Prior to the 100th transit, the curve is irregular, indicating that a number of still higher order modes are present which are damped out rapidly.

The next step in the infinite strip problem was to repeat solutions of the above type for other sets of dimensions. However, if $a^2/b\lambda$ is very small compared to $(b/a)^2$, the actual dimensions of the mirrors and their spacing are no longer important, the only parameter of importance being the Fresnel number $N = a^2/b\lambda$. This is approximately equal to the number of Fresnel zones seen in one mirror from the center of the other mirror, and as pointed out earlier, it determines the number of ripples in the field distributions. Amplitude and phase distributions for the dominant mode obtained by solving (27) are shown in Fig. 7 for different values of N. The larger the N, the weaker is the field intensity at the edge of the mirror, and the smaller is the power loss due to spill-over. The plot of power loss per transit as a function of N is approximately a straight line on log-log paper and is shown as the lowest line in Fig. 8. The phase shift per transit as a function of N is given by the lowest line in Fig. 9.

A uniform plane wave excitation can never give rise to a mode with odd symmetry. In order to investigate the possibility of modes of this type, the problem was re-programmed for an initial wave for which the field intensity over one-half the strip (0 to $+a$) was equal but opposite in sign to the field intensity over the other half of the strip (0 to $-a$). Steady-state solutions did indeed result, and odd-symmetric normal modes therefore exist. The amplitude and phase distributions are shown in Fig. 10 for several values of N. The amplitude is zero at the center, as expected. While shown for only one half of the strip, it is the same in the other half, but

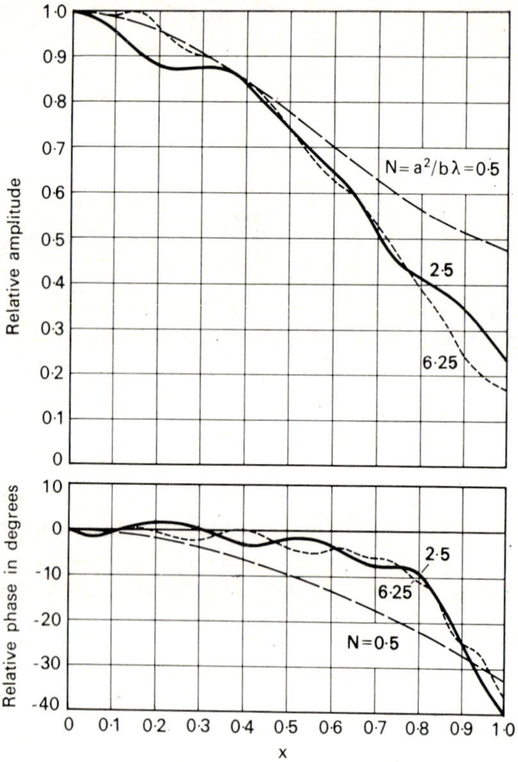

FIG. 7. Relative amplitude and phase distributions of field intensity of the lowest order even-symmetric mode for infinite strip mirrors.

with a reversal in sign. Note that for the same values of N, the amplitude at the edge is higher than for the dominant mode. The spill-over loss should be higher and this is confirmed by the loss curve in Fig. 8 labeled "infinite strip odd-symmetric mode". The corresponding phase shift curve is shown in Fig. 9.

3.3 *Circular Plane Mirrors*

The feasibility of obtaining the normal mode solutions for the infinite strip mirrors having been established, programs were next

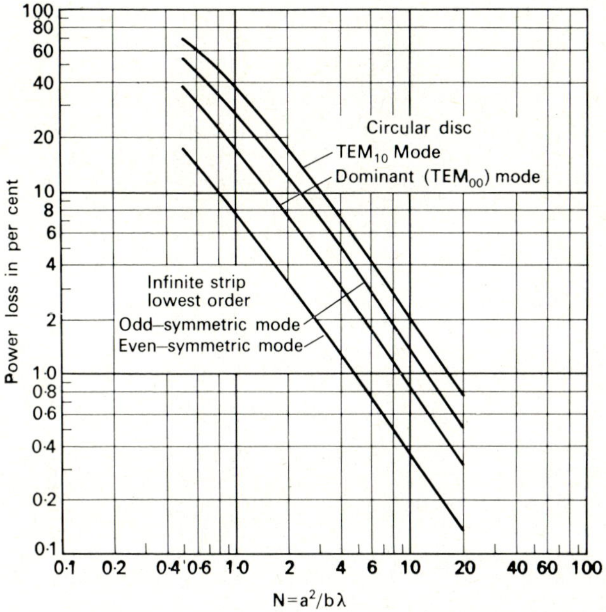

FIG. 8. Power loss per transit vs. $N = a^2/b\lambda$ for infinite strip and circular plane mirrors.

set up to investigate the modes for plane circular mirrors. The first case considered was that for uniform plane wave excitation of the system. Once again, the polarization was assumed to be everywhere parallel to the same axis, and this results in a scalar wave solution having circular symmetry [(9) with $n = 0$]. That is, the amplitude and phase of the field intensity is the same for all points at the same radius from the center. The transverse field distributions for the lowest order mode of this type are shown in Fig. 11 for various values of N. The loss and phase shift are shown in Figs. 8 and 9 under the title "circular disc (dominant mode, TEM_{00})". One hundred increments along the radius were used for the numerical integrations involved.

Next we examined modes of the odd-symmetric type for circular plane mirrors. The equation we used was (9) with $n = 1$. Fig. 12

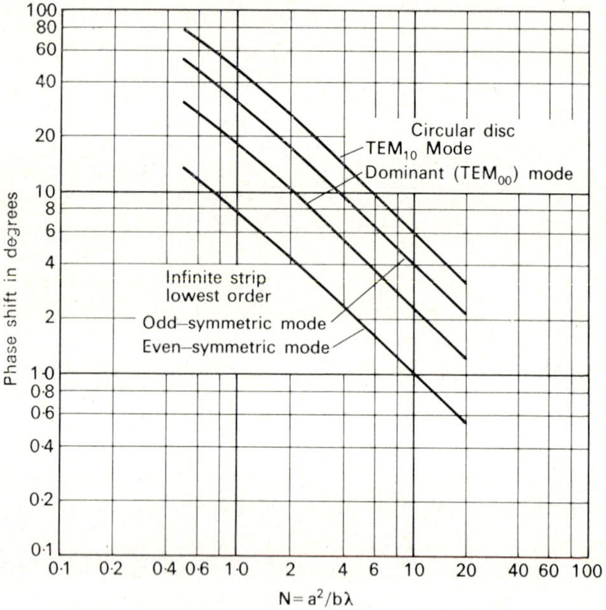

FIG. 9. Phase shift per transit (leading relative to geometrical phase shift) vs. $N = a^2/b\lambda$ for infinite strip and circular plane mirrors.

shows amplitude and phase distributions for the lowest order mode of the odd-symmetric type for circular plane mirrors. Again the loss and phase shift are given in Figs. 8 and 9 under the title "circular disc, TEM_{10} mode".

Normal modes with higher orders of angular variation ($n \geqq 2$) and radial variation ($m \geqq 1$) have greater losses and phase shifts than those of TEM_{00} and TEM_{10} modes. The mode with the least attenuation is therefore the lowest order, or TEM_{00} mode, which we designate as the dominant mode for circular plane mirrors.

3.4 *Confocal Spherical Mirrors*

Before (13) was programmed for solutions on the computer a more general method for solving the problem of the confocal

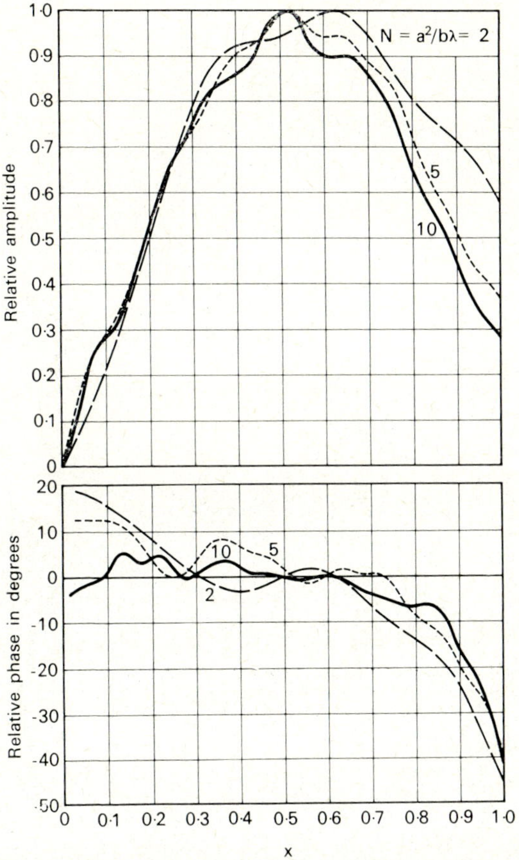

FIG. 10. Relative amplitude and phase distributions of field intensity of the lowest-order odd-symmetric mode for infinite strip mirrors.

spherical mirrors was tried—a procedure that can be used to solve problems involving mirrors with rather arbitrary but small curvatures. In this method the field at each mirror is calculated using the equation for circular plane mirrors and then a phase distribution corresponding to the curvature of the mirror is

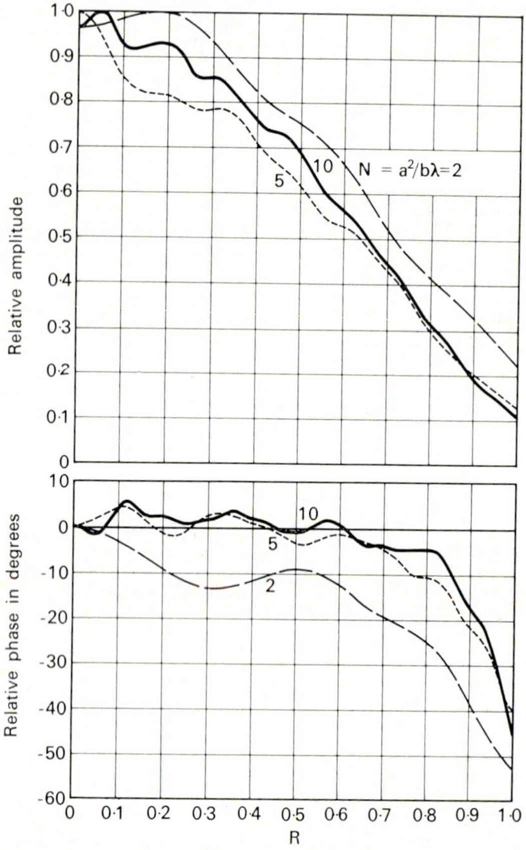

FIG. 11. Relative amplitude and phase distributions of field intensity of the dominant (TEM_{00}) mode for circular plane mirrors.

added to this field before it is used in the next iterative computation. The results from this general method of solution and from solving (13) are in perfect agreement.

The problem of confocal spherical mirrors has also been solved by Goubau[8] and Boyd and Gordon.[9] The results of their analyses are in good agreement with our computed results.

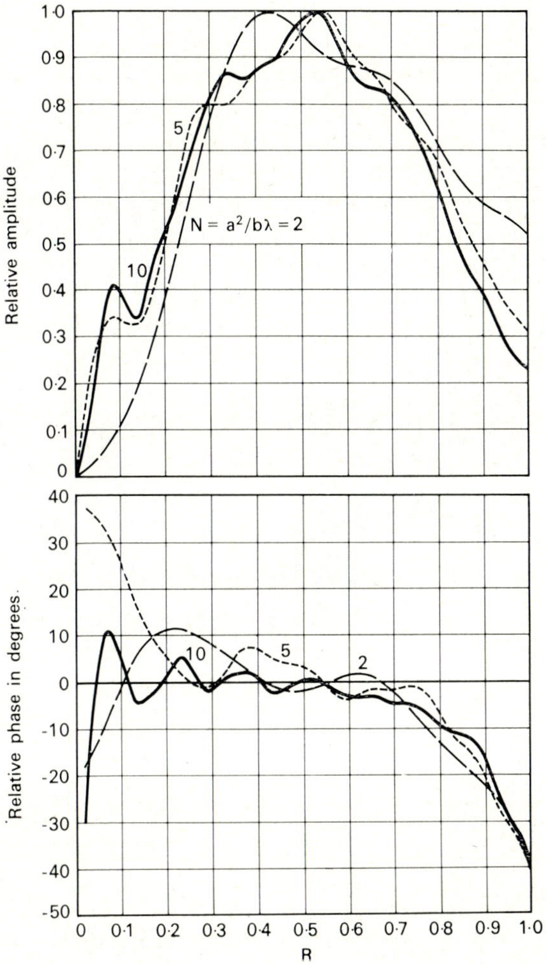

FIG. 12. Relative amplitude and phase distributions of field intensity of the TEM_{10} mode for circular plane mirrors.

Amplitude distributions of the field intensity for TEM_{00} and TEM_{10} modes are shown in Figs. 13 and 14. The phase distributions are all uniform over the surface of the mirrors and therefore are not plotted. The loss and phase shift per transit are given in Figs. 15 and 16. We note some rather remarkable differences between these solutions and those obtained for circular plane mirrors. First, the field is much more tightly concentrated near the axis of the reflector and falls to a much lower value at the edge than is true for plane mirrors; also the amplitude distribution does not have ripples in it, but is smooth. Second, the surface of the reflector coincides with the phase front of the wave, making it an equiphase surface. Third, the difference between the phase shifts for all the normal modes are integral multiples of 90 degrees. Fourth, the losses may be orders of magnitude less than those for plane mirrors.

The result that the mirror surface is an equiphase surface should not be surprising, but can be deduced from integral equation (13).

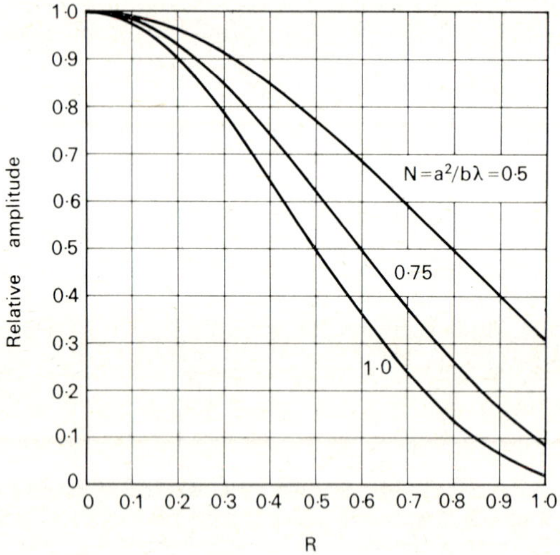

FIG. 13. Relative amplitude distribution of field intensity of the dominant (TEM_{00}) mode for confocal spherical mirrors. The relative phase distribution on the surface of the mirror is uniform.

If we associate the factor j^{n+1} with γ_n the kernel becomes real. Since the eigenvalues and eigenfunctions of a real symmetric kernel are all real,[7] we see that the field distribution is of uniform phase over the surface of the mirror. Furthermore, since $(j^{n+1}\gamma_n)$ is real, the phase shift for the normal modes belonging to a set of modes with a given angular variation must be an integral multiple of 180 degrees and the difference between the phase shifts for the normal modes with different angular variations but the same radial variation is an integral multiple of 90 degrees; that is, the phase shift is equal to $[180m + 90(n + 1)]$ degrees. Therefore, if the mirrors are adjusted for the resonance of a particular normal mode, half of the totality of all the modes are also resonant. However, the resonant mode with the lowest loss would persist longest in the resonator. Just as in the case of plane parallel mirrors, the mode with the lowest loss is the TEM_{00} mode.

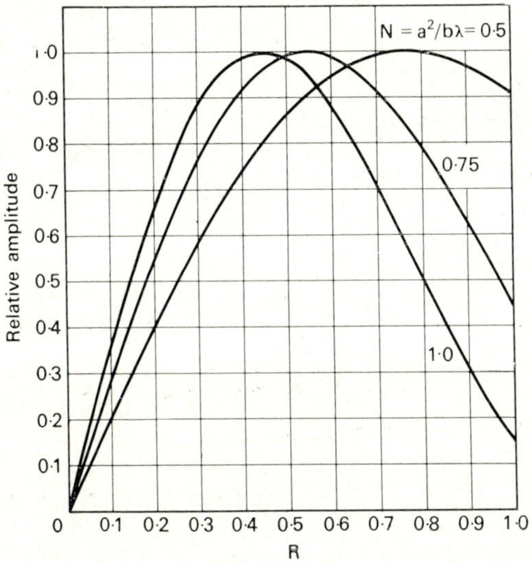

FIG. 14. Relative amplitude distribution of field intensity of the TEM_{10} mode for confocal spherical mirrors. The relative phase distribution on the surface of the mirror is uniform.

IV. Discussion of Results

The results of machine computation have shown that a two-mirror interferometer, whether of the plane or concave mirror type, can have normal modes of propagation which are self-perpetuating or self-reproducing in the distance of one transit. We use the term mode of propagation rather than mode of resonance

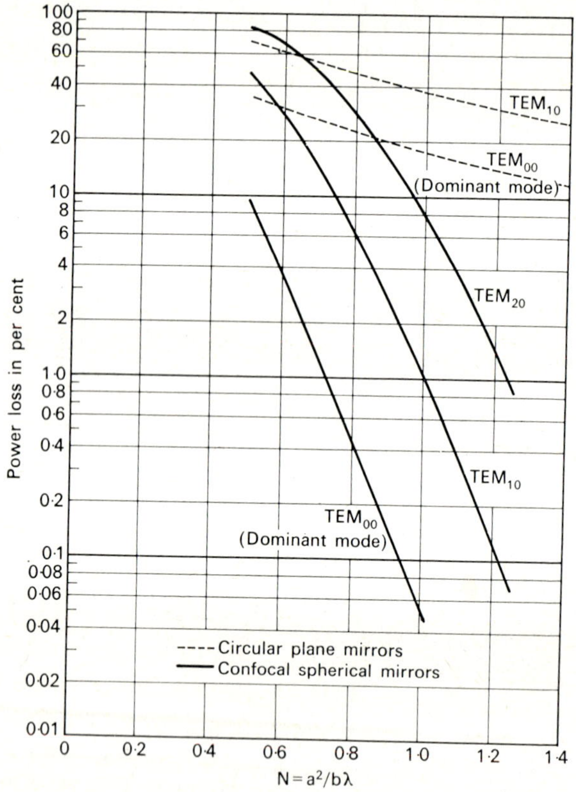

FIG. 15. Power loss per transit vs. $N = a^2/b\lambda$ for confocal spherical mirrors. (Dashed curves for circular plane mirrors are shown for comparison.)

to emphasize the fact that these steady-state solutions are the result of multiple transits whether or not the plate separation happens to be adjusted for resonance. An analog of the plane mirror interferometer is a transmission medium consisting of a series of periodic collinear apertures, as was shown in Fig. 1. The same solutions apply, and here it is clear that the reproduction of a normal mode field at successive apertures does not depend on any critical relation between b and λ.

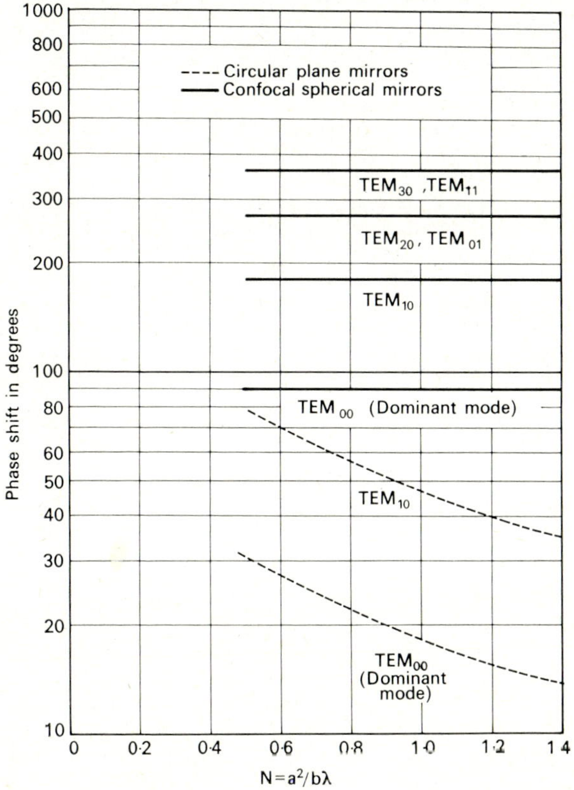

FIG. 16. Phase shift per transit (leading relative to geometrical phase shift) vs. $N = a^2/b\lambda$ for confocal spherical mirrors. (Dashed curves for circular plane mirrors are shown for comparison.)

In Fig. 17 is shown the way in which a number of square-plate modes can be synthesized from the infinite-strip modes. Diagram A shows schematically the field distribution for the dominant square-plate mode obtained as the product of the field distributions of two even-symmetric strip modes crossed at right angles and with polarization as shown. Since the eigenvalue for the square plate is the product of the eigenvalues for the two strips, the phase shift per transit is the sum of the phase shifts for the two strips and, if the loss is small, the loss per transit is essentially the sum of the losses for the two strips. Diagram B represents an odd-symmetric square-plate mode formed by taking the product of an even- and

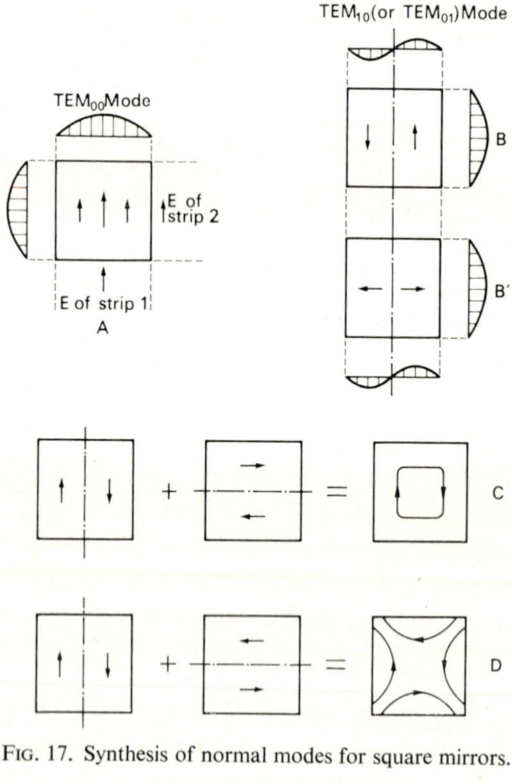

FIG. 17. Synthesis of normal modes for square mirrors.

an odd-symmetric strip mode; B′ is the same mode but with the polarization rotated 90°; C is a circular electric type of mode formed by *adding* two modes of the type B. This addition is permitted because the two components are degenerate, It follows that the circular electric mode C is degenerate with B and has the same loss and phase shift per transit. By taking the difference between the same two B modes as shown, the mode D is obtained, resembling the TE_{21} mode in circular waveguide. We give all the patterns B, B′, C and D the same designation, TEM_{10} (or TEM_{01}), since they are composites of the one basic mode type. Similar syntheses can be performed for circular mirrors, either plane or concave. It is interesting that degeneracies of this type are common for the interferometer because the electric vector E is at liberty to be parallel or perpendicular to the mirror edges. In a metal waveguide they are uncommon because the polarization of E at the boundaries is restricted.

The dominant mode and a number of higher-order modes for square and circular mirrors are depicted in Fig. 18, in which electric field vectors are shown. This classification of modes applies to plane as well as confocal spherical mirrors. In the case of

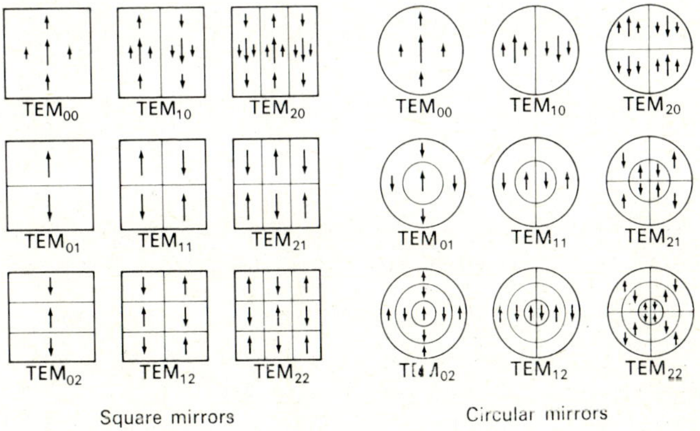

FIG. 18. Field configuration of normal modes for square and circular mirrors.

rectangular mirrors, the x axis may be taken along the longer dimension, in which case the first subscript always denotes the number of field reversals along the longer dimension.

In formulating the problem we have assumed that the waves were almost transverse electromagnetic. The solutions for the flat mirror are consistent with this assumption. At the edges of the mirror there is a phase lag of approximately 45 degrees relative to the center, but this is only one-eighth of a wavelength out of many wavelengths for the mirror diameter. Thus the curvature of the wavefront away from the transverse plane is exceedingly small, and the assumption appears justified. For higher-order modes such as B′ of Fig. 17, it is clear that the field lines must have longitudinal components. This is illustrated by an edge view in Fig. 19. However, provided the width of a cell c is much greater than a half-wavelength, the longitudinal field intensity should be negligible compared to the transverse. Only for very high-order modes should this approximation begin to fail. Because the low-order modes of importance are essentially transverse electromagnetic, they are designated as TEM modes.

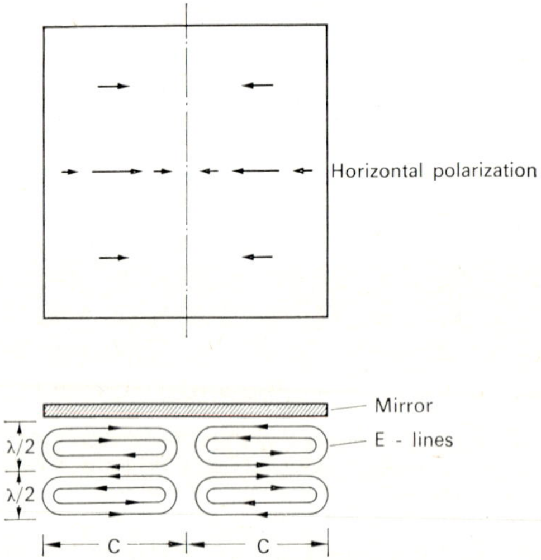

FIG. 19. Field configuration of the TEM_{10} mode for square mirrors.

The plane mirror modes have a phase which is not constant over the mirror. This does not mean that it is impossible to space the mirrors for resonance of the entire field pattern. Actually, the phase delay for one transit is the same for every point on the wavefront. Therefore, if the plates are separated by the distance b plus an additional amount for the phase shift per transit of the mode desired, that mode should resonate in the interferometer. Other modes should not be resonant for this separation because they have different phase shifts per transit.

Since the field configurations of many of the normal modes of the interferometer are very similar to those of metal tube and parallel-plane waveguides, it is not surprising to find that simple waveguide theory can be used to predict certain characteristics of the interferometer modes. One of these characteristics is phase shift per transit. For instance, the field distributions of the normal modes for infinite strip mirrors are very similar to those of the TE modes of parallel-plane waveguide; also, by adding two orthogonally polarized TEM_{10} modes for circular plane mirrors, one obtains a field configuration which is very similar to that of the circular electric (TE_{01}) mode of circular waveguide (Fig. 20). Thus the amount of phase shift per transit computed for these modes of the interferometer agrees well with the phase shifts obtained for

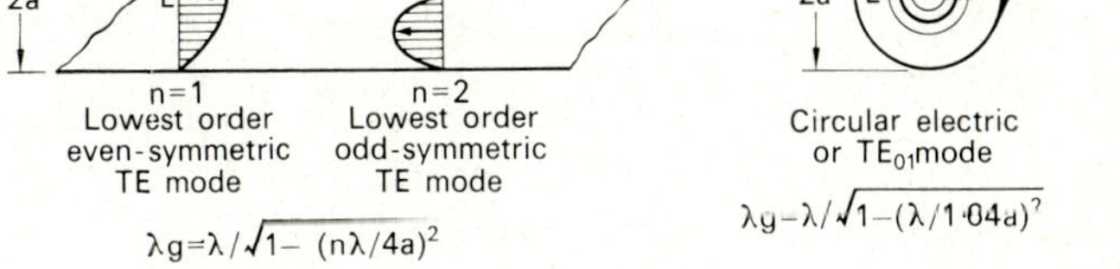

FIG. 20. TE modes in a parallel-plane waveguide and circular electric mode in a circular waveguide.

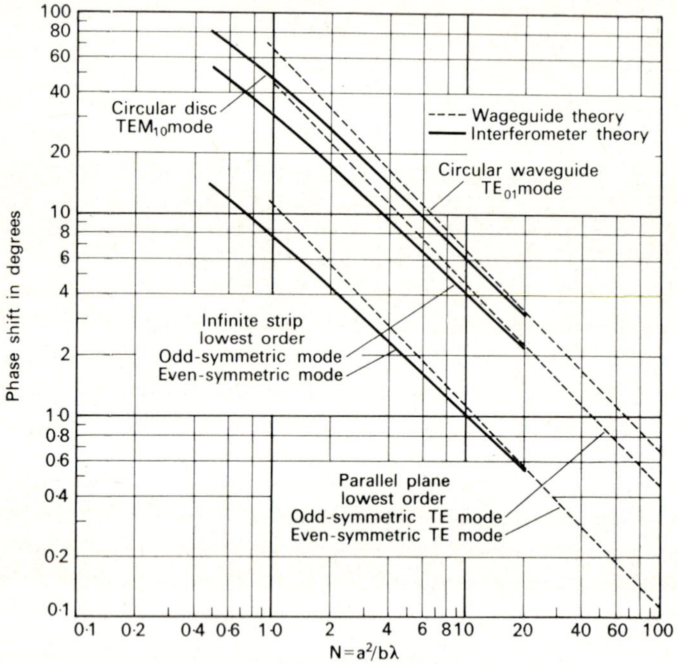

FIG. 21. Comparison of computed phase shifts based on waveguide theory and on interferometer theory.

TE modes of parallel-plane waveguide and TE_{01} mode of circular waveguide. This is illustrated in Fig. 21. We see that agreement becomes better for larger values of N. This is because the similarity between field configurations becomes closer for larger values of N.

If we regard a uniform plane wave as being resolvable into a set of normal modes, there can be no such thing as a resonance for a uniform plane wave. Why then does it appear that there is such a resonance in passive optical interferometers? It is because for the usual optical case $a^2/b\lambda$ is in the thousands. The phase shifts per transit are extremely small, hence the mode resonances lie very close together in frequency. At the same time, the reflection coefficients of the best optical mirrors are so poor, and the Q of the interferometer is so low, that the resonance line width contains hundreds of normal mode resonances. Thus the uniform plane wave undergoes very little decomposition when resonated. Nevertheless, in the case of an active interferometer, the decomposition may be complete.

We now make use of the formula for the Q of a resonant waveguide cavity to compute the Q of an interferometer system. The Q of a resonant waveguide cavity is given by

$$Q = \frac{|R_1 R_2 e^{-2\alpha b}|}{1 - |R_1 R_2 e^{-2\alpha b}|} \left(\frac{2\pi b}{\lambda_g}\right)\left(\frac{\lambda_g}{\lambda}\right)^2, \tag{14}$$

where α is the attenuation constant of the waveguide and λ_g is the guide wavelength. For the interferometer we assume that α is zero and that λ_g is equal to λ, the free space wavelength. The voltage reflection coefficients R_1 and R_2 for the two reflectors are given by

$$|R_1| = |R_2| = \sqrt{1 - \delta_r - \delta_d}, \tag{15}$$

where δ_r is the power loss in reflection and δ_d is the power loss due to spill-over. When these losses are very small, Q reduces to

$$Q \cong \frac{1}{\delta_r + \delta_d} 2\pi \frac{b}{\lambda}. \tag{16}$$

Hence

$$\frac{1}{Q} = \frac{1}{Q_r} + \frac{1}{Q_d}, \tag{17}$$

where

$$Q_r = \frac{2\pi b}{\lambda \, \delta_r}, \qquad Q_d = \frac{2\pi b}{\lambda \, \delta_d}. \tag{18}$$

The resonance line width at half-power points given as the change in electrical length of the resonator, $\Delta\phi$, is

$$\begin{aligned} \Delta\phi &= 2\pi \left(\frac{b}{\lambda}\right)\left(\frac{\Delta\lambda}{\lambda}\right) \\ &= \delta_r + \delta_d \text{ radians}, \end{aligned} \tag{19}$$

where we have substituted $(1/Q)$ for $(\Delta\lambda/\lambda)$.

Let us consider an interferometer having circular plane mirrors with $2a = 1$ cm, $b = 20$ cm, $\lambda = 5 \times 10^{-5}$ cm and a reflection loss of $\delta_r = 0.02$. In this case $N = a^2/b\lambda = 250$. Extrapolating the loss and phase shift curves of Figs. 8 and 9, we obtain diffraction loss $\delta_d = 9 \times 10^{-5}$ and phase shift for the dominant (TEM_{00}) mode $\phi_d = 0.11$ degree. The diffraction loss is thus negligible compared to reflection loss, which limits the Q to a value of 1.25×10^8. The phase shift for the next higher order (TEM_{10}) mode is 0.30 degree and therefore it is separated from the dominant (TEM_{00}) mode by 0.19 degree or 0.0033 radian. The resonance line width, as given by (19), is 0.02 radian. Thus we see that TEM_{00} and TEM_{10} modes are not resolved. As the mirror separation is reduced or mirror size increased, more and more normal modes will become unresolved and a uniform plane wave will suffer less decomposition when resonated.

When an interferometer is filled with an active medium, the medium can compensate for the mirror losses and yield an enormously increased Q. Under these circumstances, the modes may be clearly resolved, and their Q's will be determined by the diffraction losses. If the gain of the medium is increased until it compensates for mirror losses plus the diffraction loss of the lowest order mode, that mode will become unstable and oscillation can result. All higher-order modes will be stable and have positive net loss. If the gain of the medium is further increased, then many modes may become unstable. In starting from a quiescent condition, spontaneous emission can initiate a large number of characteristic waves in the interferometer. These may then start to grow, but the dominant mode will always grow faster and should saturate first. At saturation the steady-state field distribution will be considerably altered. The relative field at the edges of the mirrors should increase, thereby increasing the relative power loss. This can be described as a coupling of power into other modes as a result of the nonlinearity of the medium. No attempt has yet been made to analyze this situation. The linear theory is at present of most interest because it allows the computation of the starting conditions for oscillation.

With the development of the normal mode picture of interferometer operation and the computation of the losses for these modes, we may now ask if there is an optimum geometry for a maser interferometer which will permit oscillation for the lowest possible gain in the medium. We know that the power gained from the medium can be increased by increasing length. For very great lengths corresponding to the far-field region ($N < 0.1$), the power gained from the medium increases more rapidly than transmission loss as length is increased, and there must always be some length beyond which oscillations can occur. However, these lengths are too great to be of practical interest. In the near-field region ($N > 1$), represented by the curves of Fig. 8, the diffraction loss increases more rapidly than the medium gain. Therefore, if the reflection loss is sufficiently small, an optimum length may exist which is most favorable for oscillation.

To be more specific, let us consider a circular plane mirror interferometer. Fom Fig. 8 we find that the loss for the dominant mode may be represented by the expression

$$\delta_d = 0.207\left(\frac{b\lambda}{a^2}\right)^{1.4}. \tag{20}$$

In order to find the optimum value of b to give a maximum Q, (20) and (18) are substituted in (17) and the resulting equation is differentiated with respect to b. For the optimum b, the diffraction loss is 2.5 times the reflection loss, and not equal to it, as might be supposed. Moreover, this result is general and holds for all modes and all shapes of plane mirrors represented in Fig. 8, provided the optimum falls on the straight-line portions of the loss curves. Since the power supplied by the medium is proportional to the stored energy in the interferometer, while the power loss of the passive interferometer is just ω/Q times the stored energy, oscillation is most likely to occur when Q is a maximum. Fig. 22 illustrates the way the interferometer dimensions affect Q. If a given mirror diameter is chosen (as represented by the dashed line A), there is clearly an optimum distance b which will produce a maximum Q (intersection of lines A and B). However, if the

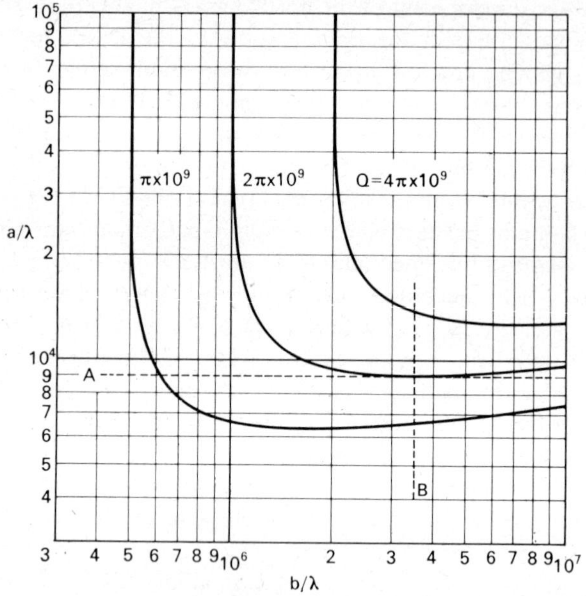

FIG. 22. Interferometer dimensions for constant Q. (Circular plane mirrors, reflection loss $= \delta_r = 0.001$.)

distance b is held constant, there is no optimum value for a. The larger a, the higher will be Q, although it will approach a limiting value beyond which there is nothing to be gained by further increase of a.

As an example, let us assume a case where

$$\lambda = 10^{-4} \text{ cm},$$

$$2a = \text{plate diameter} = 2 \text{ cm},$$

$$\delta_r = \text{power reflection loss} = 0.001.$$

The optimum proportions require that δ_d be 0.0025, and for this, b is 435 cm and the resulting Q is 7.8×10^9. The length of 435 cm is probably impractically large for a maser. If b is reduced to a more reasonable value of 50 cm, the Q will drop to 3.14×10^9, which is the limiting value due to reflection loss. (The value assumed here for δ_r is already much lower than can be obtained from evaporated metal films and would require the technique of multilayered dielectric films.) In order to oscillate, the active medium would have to have a power amplification factor in excess of 1.00002 per centimeter of path.

In the case of confocal paraboloidal mirrors of 2 cm diameter, the optimum length turns out to be 8900 cm. If the diameter is reduced to 0.5 cm, the optimum length is still 530 cm, and for these proportions Q is 3.1×10^{10}. It is clear that with confocal mirrors the diffraction losses are negligible for any reasonable proportions of the interferometer.

One question of importance is whether there is an optimum set of dimensions which will discriminate against unwanted modes. It has sometimes been suggested that by making the mirror diameter small relative to the mirror spacing, "slant rays" will be more rapidly lost from the system. However, from Fig. 8 it can be seen that the ratios of the losses for the several modes is independent of N provided N is greater than 1. Thus, if diffraction losses predominate, there is no way of discriminating against unwanted modes by juggling dimensions. The limiting amount of discrimination is merely governed by the ratio of the losses for the different modes, which is independent of the dimensions. However, if reflection losses predominate, the discrimination between lower-order modes would be almost nonexistent and it would be advantageous to increase mirror separation and/or decrease mirror dimensions so as to make diffraction losses predominate. In the case of the confocal mirrors, the loss ratios between modes are not constant (Fig 15) although, for values of N larger than those shown, they may become so. At any rate, for values of N close to unity, a small amount of increased discrimination against higher order modes can be obtained by making the mirrors *larger*.

V. Conclusions

Diffraction studies carried out on the IBM computer have led to the following conclusions.

1. Fabry–Perot interferometers, whether of the plane or concave mirror type, are characterized by a discrete set of normal modes which can be defined on an iterative basis. The dominant mode has a field intensity which falls to low values at the edges of the mirrors, thereby causing the power loss due to diffractive spillover to be much lower than would be predicted on the assumption of uniform plane wave excitation.

2. Uniform plane waves are not normal modes for a flat-plate interferometer. Consequently, interferometer resonances do not exist for "slant rays," i.e., plane waves traveling at an angle with respect to the longitudinal axis.

3. The losses for the dominant mode of the plane mirror system are so low that for most practical geometries performance will be limited by reflection losses and scattering due to aberrations. For confocal mirrors the diffraction losses are even lower.

4. There are no higher-order modes with losses lower than the dominant (lowest-order) mode.

5. The ratio of diffraction losses between the modes investigated for the plane mirror system is independent of the interferometer dimensions in the range of interest. Therefore, if diffraction losses predominate, there is no way of proportioning the interferometer so as to favor any one mode.

The computer technique we employed is general and versatile. It can be used for studying mirrors having rather arbitrary but small curvatures. With little modification, the same technique can be used to study the effects of aberration and misalignment.

Appendix A

Rectangular Plane Mirrors

The geometry for rectangular plane mirrors parallel to the xy plane is shown in Fig. 2. According to (1), the iterative equation for computing the field at the surface of mirrors is

$$u_{q+1}(x_2, y_2) = \frac{j}{2\lambda}\int_{-c}^{c}\int_{-a}^{a} u_q(x_1, y_1)\frac{e^{-jkR}}{R}\left(1 + \frac{b}{R}\right) dx_1\, dy_1, \tag{21}$$

where

$$R = \sqrt{b^2 + (x_1 - x_2)^2 + (y_1 - y_2)^2}.$$

If b/a and b/c are large, (21) can be reduced to

$$u_{q+1}(x_2, y_2) = \frac{je^{-jkb}}{\lambda b}$$

$$\times \int_{-c}^{c} \int_{-a}^{a} u_q(x_1, y_1) e^{-jk[(x_1-x_2)^2+(y_1-y_2)^2]/2b} \, dx_1 \, dy_1, \qquad (22)$$

which is valid for $(a^2/b\lambda) \ll (b/a)^2$ and $(c^2/b\lambda) \ll (b/c)^2$.* The corresponding integral equation is

$$v(x_2, y_2) = \gamma \int_{-c}^{c} \int_{-a}^{a} K(x_2, x_1; y_2, y_1) v(x_1, y_1) \, dx_1 \, dy_1, \qquad (23)$$

where

$$K(x_2, x_1; y_2, y_1) = \frac{j}{\lambda b} e^{-jk[(x_1-x_2)^2+(y_1-y_2)^2]/2b} \qquad (23a)$$

and the factor e^{-jkb} is absorbed in γ.

Here, the kernel of the integral equation is separable in x and y. If the distribution function v is assumed to be of the form

$$v(x, y) = v_x(x) v_y(y) \qquad (24)$$

it is possible to separate (23) into two equations, one involving x only and the other involving y only; that is,

$$v_x(x_2) = \gamma_x \int_{-a}^{a} K_x(x_2, x_1) v_x(x_1) \, dx_1, \qquad (25a)$$

$$v_y(y_2) = \gamma_y \int_{-c}^{c} K_y(y_2, y_1) v_y(y_1) \, dy_1, \qquad (25b)$$

* Actually, the stringency of this requirement can be relaxed somewhat for lower-order modes in which field intensities near the edges of the mirror are rather low. We have made check computations for the case $a^2/b\lambda = 5$ and $(b/a)^2 = 25$ and have found that the results based on the exact equation and on the approximate equation are in essential agreement.

with

$$K_x = \frac{e^{j(\pi/4)}}{\sqrt{\lambda b}} e^{-jk(x_1 - x_2)^2/2b}, \tag{25c}$$

and

$$K_y = \frac{e^{j(\pi/4)}}{\sqrt{\lambda b}} e^{-jk(y_1 - y_2)^2/2b}. \tag{25d}$$

The product of the eigenvalues γ_x and γ_y is equal to the eigenvalue γ in (23).

It remains to be shown that (25a) through (25d) represent integral equations for infinite strip mirrors. Let us consider a pair of infinite strip mirrors of width $2a$ and separated by b. The iterative equation for computing the field at the mirrors can be derived from (1). It is

$$u_{q+1}(x_2) = \frac{e^{j(\pi/4)}}{2\sqrt{\lambda}} \int_{-a}^{a} u_q(x_1) \frac{e^{-jk\rho}}{\sqrt{\rho}} \left(1 + \frac{b}{\rho}\right) dx_1, \tag{26}$$

where

$$\rho = \sqrt{b^2 + (x_1 - x_2)^2}.$$

For $(a^2/b\lambda) \ll (b/a)^2$, (26) reduces to

$$u_{q+1}(x_2) = \frac{e^{j[(\pi/4) - kb]}}{\sqrt{\lambda b}} \int_{-a}^{a} u_q(x_1) e^{-jk(x_1 - x_2)^2/2b} \, dx_1. \tag{27}$$

The corresponding integral equation is

$$v(x_2) = \gamma \int_{-a}^{a} K(x_2, x_1) \, v(x_1) \, dx_1, \tag{28}$$

where

$$K(x_2, x_1) = \frac{e^{j(\pi/4)}}{\sqrt{\lambda b}} e^{-jk(x_1 - x_2)^2/2b} \tag{28a}$$

and the factor e^{-jkb} is absorbed in γ. We see that (25) and (28) are identical in form.

Appendix B

Circular Plane Mirrors

Assuming approximately plane waves propagating normally to the circular plane mirrors (Fig. 3), the iterative equation for computing the steady-state field distribution can be written as

$$u_{q+1}(r_2, \phi_2) = \frac{j}{2\lambda}\int_0^a \int_0^{2\pi} u_q(r_1, \phi_1)\frac{e^{-jkR}}{R}\left(1 + \frac{b}{R}\right) r_1\, d\phi_1\, dr_1, \qquad (29)$$

where

$$R = \sqrt{b^2 + {r_1}^2 + {r_2}^2 - 2r_1 r_2 \cos(\phi_1 - \phi_2)}.$$

If b/a is large, (29) simplifies to

$$u_{q+1}(r_2, \phi_2) = \frac{je^{-jkb}}{\lambda b}\int_0^a \int_0^{2\pi} u_q(r_1, \phi_1) \cdot e^{-jk[(r_1{}^2 + r_2{}^2)/2b - (r_1 r_2/b)\cos(\phi_1 - \phi_2)]} r_1\, d\phi_1\, dr_1, \qquad (30)$$

which is valid for $(a^2/b\lambda) \ll (b/a)^2$.*

The integral equation corresponding to (30) is

$$v(r_2, \phi_2) = \gamma \int_0^a \int_0^{2\pi} K(r_2, \phi_2; r_1, \phi_1) v(r_1, \phi_1) r_1\, d\phi_1\, dr_1, \qquad (31)$$

with

$$K(r_2, \phi_2; r_1, \phi_1) = \frac{j}{\lambda b} e^{-jk[(r_1{}^2 + r_2{}^2)/2b - (r_1 r_2/b)\cos(\phi_1 - \phi_2)]} \qquad (31a)$$

and where the factor e^{-jkb} is absorbed in γ. Making use of the relation[10]

$$e^{jn[(\pi/2) - \phi_2]} J_n\left(k\frac{r_1 r_2}{b}\right) = \frac{1}{2\pi}\int_0^{2\pi} e^{jk(r_1 r_2/b)\cos(\phi_1 - \phi_2) - jn\phi_1}\, d\phi_1 \qquad (32)$$

* Comments in Appendix A regarding the stringency of this requirement are also applicable herein.

and integrating (31) with respect to ϕ_1, it is seen that

$$v(r, \phi) = R_n(r)e^{-jn\phi}, \qquad (n = \text{integer}) \tag{33}$$

satisfies (31). The function $R_n(r)$ satisfies the reduced integral equation

$$R_n(r_2)\sqrt{r_2} = \gamma_n \int_0^a K_n(r_2, r_1)R_n(r_1)\sqrt{r_1}\, dr_1, \tag{34}$$

with

$$K_n(r_2, r_1) = \frac{j^{n+1}k}{b} J_n\left(k\frac{r_1 r_2}{b}\right)\sqrt{r_1 r_2}\, e^{-jk(r_1{}^2 + r_2{}^2)/2b}, \tag{34a}$$

where J_n is a Bessel function of the first kind and nth order.

Appendix C

Confocal Spherical or Paraboloidal Mirrors

For confocal spherical mirrors of circular cross section (Fig. 4), the iterative equation corresponding to (29) is

$$u_{q+1}(r_2, \phi_2) = \frac{j}{2\lambda}\int_0^a\int_0^{2\pi} u_q(r_1, \phi_1)\frac{e^{-jkR}}{R}\left(1 + \frac{b_1}{R}\right) r_1\, d\phi_1\, dr_1 \tag{35}$$

where

$$R = \sqrt{b_1{}^2 + r_1{}^2 + r^2_2 - 2r_1 r_2 \cos(\phi_1 - \phi_2)}.$$

The distance b_1 is given by

$$b_1 = b - \Delta_1 - \Delta_2 \tag{36}$$

where, for confocal spherical mirrors,

$$\Delta_i = b - \sqrt{b^2 - r_i^2} \qquad i = 1, 2. \tag{36a}$$

If b/a is large, the distance Δ_i is given approximately by

$$\Delta_i \cong r_i{}^2/2b \qquad i = 1, 2, \tag{37}$$

which is exact for confocal paraboloids. In this case (35) simplifies to

$$u_{q+1}(r_2,\phi_2) = \frac{je^{-jkb}}{\lambda b}\int_0^a\int_0^{2\pi} u_q(r_1,\phi_1)\, e^{jk(r_1 r_2/b)\cos(\phi_1-\phi_2)} r_1\, d\phi_1\, dr_1, \tag{38}$$

which is valid for $(a^2/b\lambda) \ll (b/a)^2$.

The integral equation corresponding to (38) is

$$v(r_2,\phi_2) = \gamma \int_0^a\int_0^{2\pi} K(r_2,\phi_2;\, r_1,\phi_1) v(r_1,\phi_1) r_1\, d\phi_1\, dr_1, \tag{39}$$

with

$$K(r_2,\phi_2;r_1,\phi_1) = \frac{j}{\lambda b} e^{jk(r_1 r_2/b)\cos(\phi_1-\phi_2)} \tag{40}$$

and where the factor e^{-jkb} is absorbed in γ. Just as in the case of circular plane mirrors, it can be shown that

$$v(r,\phi) = S_n(r)e^{-jn\phi} \qquad (n = \text{integer}) \tag{41}$$

satisfies (39). The function $S_n(r)$ satisfies the reduced integral equation

$$S_n(r_2)\sqrt{r_2} = \gamma_n \int_0^a K_n(r_2,r_1) S_n(r_1)\sqrt{r_1}\, dr_1, \tag{42}$$

with

$$K_n(r_2,r_1) = \frac{j^{n+1}k}{b} J_n\left(k\frac{r_1 r_2}{b}\right)\sqrt{r_1 r_2}. \tag{42a}$$

References

1. Schawlow, A. L. and Townes, C. H., Infared and Optical Masers, *Phys. Rev.*, **112,** 1958, p. 1940.
2. Maiman, T. H., Stimulated Optical Radiation in Ruby, *Nature*, **187,** 1960, p. 493.
3. Collins, R. J., Nelson, D. F., Schawlow, A. L., Bond, W., Garrett, C. G. B. and Kaiser, W., Coherence, Narrowing, Directionality and Relaxation Oscillations in the Light Emission from Ruby, *Phys. Rev. Letters*, **5,** 1960, p. 303.

4. SILVER, S., *Microwave Antenna Theory and Design*, McGraw-Hill, New York, 1949, p. 167.
5. HILDEBRAND, F. B., *Methods of Applied Mathematics*, Prentice-Hall, Englewood Cliffs, N.J., 1952.
6. CONNES, P., Increase of the Product of Luminosity and Resolving Power of Interferometers by Using a Path Difference Independent of the Angle of Incidence, *Revue d'Optique*, **35,** 1956, p. 37.
7. LOVITT, W. V., *Linear Intergral Equations*, Dover Publications, New York, 1950, pp. 129; 137.
8. GOUBAU, G. and SCHWERING, F., On the Guided Propagation of Electromagnetic Wave Beams, URSI-IRE Spring Meeting, May 1960, Washington, D.C.
9. BOYD, G. D. and GORDON, J. P., Confocal Multimode Resonator for Millimeter through Optical Wavelength Masers, *Bell Syst. Tech. J.* **40**, **489**.
10. STRATTON, J. A., *Electromagnetic Theory*, McGraw-Hill, New York, 1941, p. 372.

4

Optical and Microwave-optical Experiments in Ruby*

T. H. MAIMAN

Hughes Research Laboratories, Malibu, California

SEVERAL recent papers[1-4] have reported optical and microwave-optical measurements in ruby (Cr^{+++} in Al_2O_3). We wish to report here some new experiments concerning the fluorescent relaxation processes in this crystal. Reported here also are the first observations of ground-state population changes in ruby due to optical excitation and the detection of optical absorption between two excited states in this crystal.

The predominant processes which ensue in a fluorescent material when it is irradiated at an appropriate wavelength are shown in Fig. 1. W_{13} is the induced transition probability per unit time due to an exciting radiation and the S_{mn} are decay rates which include both radiative and non-radiative processes. In this crystal S_{21} is easily obtained from the decay rate of the fluorescent level (^2E) after an exciting source is turned off. The lifetime for this process is about 5 msec. Varsanyi, Wood, and Schawlow[2] have further demonstrated that this lifetime is almost entirely due to spontaneous emission, i.e., S_{21} is approximately the Einstein A coefficient A_{21}.

An approximate value for the rate S_{32} was obtained in the following way. A crystal of ruby was irradiated with 5600 Å radiation

Phys. Rev. Letters* **4, 564 (1960).

causing absorption into the lower band ($^4A_2 \rightarrow {}^4F_2$). The sample used was a one-centimeter cube cut from a boule of standard pink ruby supplied by the Linde Company, with a concentration of approximately 0.05 weight percent of Cr_2O_3 to Al_2O_3. Two components of radiation re-emitted from the crystal were observed in a direction perpendicular to the exciting beam: that due to re-emission of the incoming radiation (spontaneous decay from $^4F_2$) and

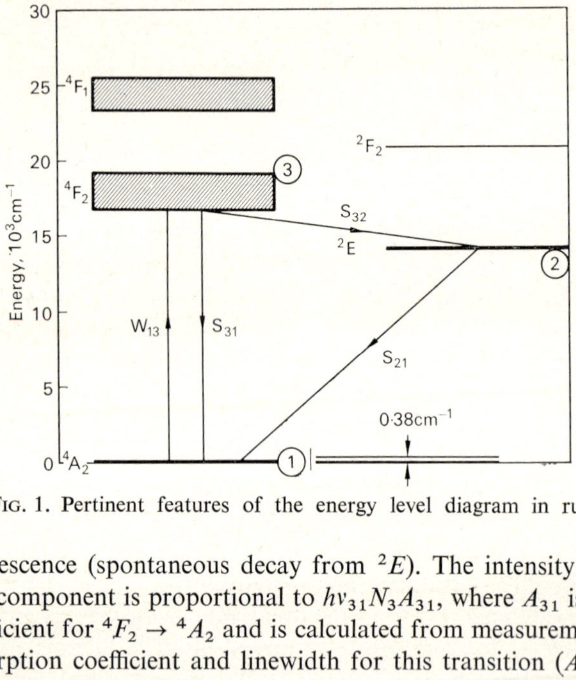

FIG. 1. Pertinent features of the energy level diagram in ruby.

fluorescence (spontaneous decay from ^2E). The intensity of the first component is proportional to $h\nu_{31}N_3A_{31}$, where A_{31} is the A coefficient for $^4F_2 \rightarrow {}^4A_2$ and is calculated from measurements of absorption coefficient and linewidth for this transition ($A_{31} \sim 3 \times 10^5$/sec). Similarly the fluorescent intensity is proportional to $h\nu_{21}N_2A_{21}$. By a measurement of the ratio of these two components and the use of an auxiliary condition applicable to steady-state conditions $N_2S_{21} = N_3S_{32}$ and also the use of the approximation $S_{21} = A_{21}$, we find $S_{32} \sim 2 \times 10^7$/sec.

A measurement of fluorescent quantum efficiency, i.e., the number of fluorescent quanta emitted compared to the number absorbed by the crystal from the exciting beam, yielded a value near unity. This result reconfirms the evidence that the life of level 2 is near radiative and also implies the $S_{32} \gg S_{31}$. The experiment was not accurate enough to yield a precise value but does indicate that the nonradiative process $(S_{31} - A_{31}) < 4 \times 10^6/\text{sec}$.

Calculations utilizing the previous results indicated that population changes in the ground state of ruby due to optical excitation would be easily observed. This conclusion was verified in the following experiments. A ruby crystal was mounted between parallel silvered plates to form a microwave cavity resonant at the ground-state zero-field splitting (11.3 kMc/sec). About half the cavity losses were due to magnetic absorption as evidenced by an increase in cavity Q when a small magnet was brought near the ruby. The reflection coefficient of the cavity was monitored on an oscilloscope while a short pulse (200 μsec) of light from a flash tube irradiated the crystal. The magnitude of the microwave magnetic absorption was observed to decrease abruptly and then return to equilibrium with a time constant of about 5 msec (see Fig. 2). We attribute this effect to temporary depletion of this ground state population with subsequent decay back from the fluorescent level. The experiment was performed at room temperature where the thermal relaxation times in the ground state of ruby are the order of a microsecond; in the time scale of the experiment, therefore, Boltzmann equilibrium in these levels is maintained.

A repetition of the above experiment at liquid helium temperatures is being planned. At this temperature we would expect to be able to observe directly any preferential depopulation of the ground sublevels due to polarized light and also any preferential repopulation of these levels since the thermal relaxation times would then be 30–100 msec.

To verify further the depletion of ground-state population observed in the previous experiment an independent measurement was made. A beam of monochromatic light of wavelength 4100 Å was transmitted through a ruby crystal and partially absorbed due to the transition ${}^4A_2 \rightarrow {}^4F_1$. When the intense pulse of radiation at 5600 Å was turned on, the 4100 Å radiation passing through the

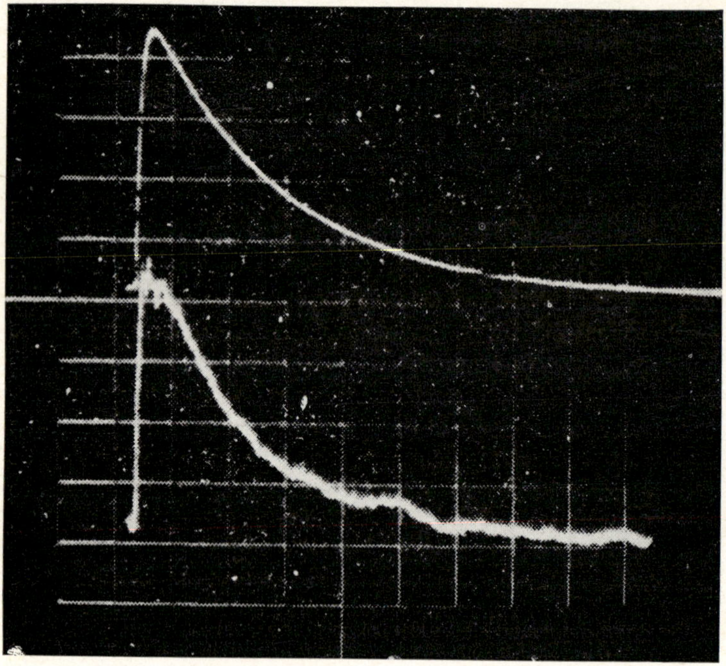

FIG. 2. The upper trace was obtained from a photomultiplier monitoring the R-line fluorescence ($^2E \rightarrow {}^4A_2$) in ruby. The lower trace is a recording of the reflection coefficient of a microwave ruby cavity during the same time interval. The time scale is 2 milliseconds per division. Both recordings followed an intense light pulse which excited the crystal via $^4A_2 \rightarrow {}^4F_2$.

crystal abruptly increased and subsequently decayed in about 5 msec just as the microwave signal in the previous experiment. This result was expected since the temporary reduction in ground-state population caused the crystal to become more transparent to the 4100 Å radiation until the fluorescent level decayed to normal. In both experiments a population change of about 3% was estimated.

An unexpected result was observed when the probe wavelength was changed from 4100 Å to 3600 Å. In this case a *decrease* in light intensity emerging from the crystal was observed. This implies that the crystal became more absorbing even though the ground-state population was decreased. We can explain this last effect when it is realized that 3600 Å radiation can cause transitions from the fluorescent level (^2E) to a high-lying charge transfer band (not shown in the figure). Consequently, we conclude that we were observing transitions between two excited optical states. The fact that the abruptly increased 3600 Å absorption also decayed with a 5-msec time constant is consistent with and strengthens the above conclusion.

The author gratefully acknowledges helpful discussions with G. Birnbaum, R. W. Hellwarth, L. C. Levitt, and R. A. Satten and is particularly indebted to I. J. D'Haenens for technical assistance in performing the experiments.

References

1. I. Wieder, *Phys. Rev. Letters* **3,** 468 (1959).
2. F. Varsanyi, D. L. Wood, and A. L. Schawlow, *Phys. Rev. Letters* **3,** 544 (1959).
3. S. Geschwind, R. J. Collins, and A. L. Schawlow, *Phys. Rev. Letters* **3,** 545 (1959).
4. J. Brossel, S. Geschwind, and A. L. Schawlow, *Phys. Rev. Letters* **3,** 548 (1959).

5

Stimulated Optical Radiation in Ruby*

T. H. MAIMAN

Hughes Research Laboratories, A Division of Hughes Aircraft Co., Malibu, California

SCHAWLOW and Townes[1] have proposed a technique for the generation of very monochromatic radiation in the infra-red optical region of the spectrum using an alkali vapour as the active medium. Javan[2] and Sanders[3] have discussed proposals involving electron-excited gaseous systems. In this laboratory an optical pumping technique has been successfully applied to a fluorescent solid resulting in the attainment of negative temperatures and stimulated optical emission at a wave-length of 6943 Å; the active material used was ruby (chromium in corundum).

A simplified energy-level diagram for triply ionized chromium in this crystal is shown in Fig. 1. When this material is irradiated with energy at a wave-length of about 5500 Å, chromium ions are excited to the $^4F_2$ state and then quickly lose some of their excitation energy through non-radiative transitions to the ^2E state[4]. This state then slowly decays by spontaneously emitting a sharp doublet the components of which at 300°K are at 6943 Å and 6929 Å. (Fig. 2a). Under very intense excitation the population of this metastable state (^2E) can become greater than that of the

* *Nature* **187**, 493 (1960).

[1]Schawlow, A. L., and Townes, C. H., *Phys. Rev.*, **112,** 1940 (1958).
[2]Javan, A., *Phys. Rev. Letters*, **3,** 87 (1959).
[3]Sanders, J. H., *Phys. Rev. Letters*, **3,** 86 (1959).
[4]Maiman, T. H., *Phys. Rev. Letters*, **4,** 564 (1960).

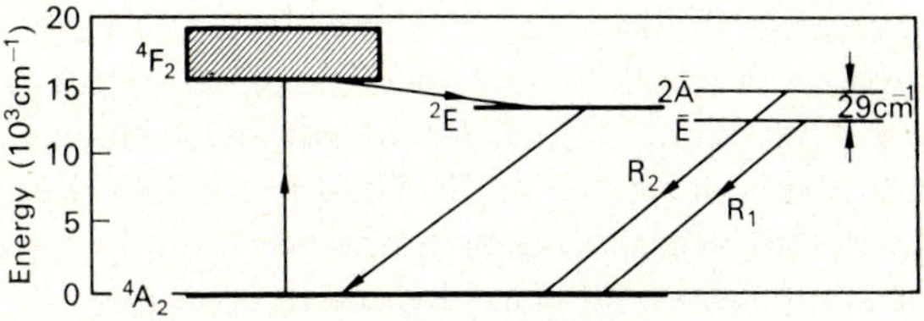

FIG. 1. Energy-level diagram of Cr^{3+} in corundum, showing pertinent processes.

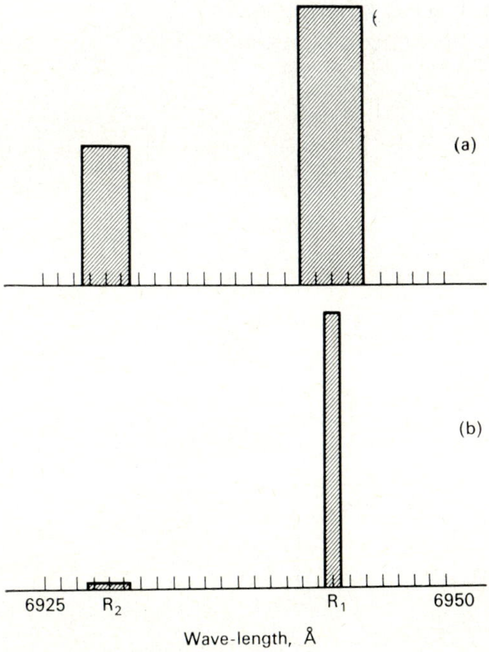

FIG. 2. Emission spectrum of ruby: (a) low-power excitation; (b) high-power excitation.

ground-state; this is the condition for negative temperatures and consequently amplification via stimulated emission.

To demonstrate the above effect a ruby crystal of 1-cm. dimensions coated on two parallel faces with silver was irradiated by a high-power flash lamp; the emission spectrum obtained under these conditions is shown in Fig. 2b. These results can be explained on the basis that negative temperatures were produced and regenerative amplification ensued. I expect, in principle, a considerably greater ($\sim 10^8$) reduction in line width when mode selection techniques are used.[1]

I gratefully acknowledge helpful discussions with G. Birnbaum, R. W. Hellwarth, L. C. Levitt, and R. A. Satten and am indebted to I. J. D'Haenens and C. K. Asawa for technical assistance in obtaining the measurements.

[1] Schawlow, A. L. and Townes, C. H., *Phys. Rev.*, **112,** 1940 (1958).

6

Stimulated Infrared Emission from Trivalent Uranium*

P. P. SOROKIN AND M. J. STEVENSON

Research Center, International Business Machines Corporation, Yorktown, New York

SCHAWLOW and Townes[1] have proposed the extension of maser techniques to generate coherent light in the visible and infrared regions of the electromagnetic spectrum. Recently Maiman[2] and Collins *et al.*[3] obtained stimulated emission of radiation in chromium-doped Al_2O_3 in the R_1 transition from the ^2E metastable state to the ground state. The oscillation was achieved by inverting the population of the metastable state relative to the ground state by a sufficient amount to overcome the losses in the optical cavity. Since this required approximately 10^{18}–10^{19} ions in the metastable state, very high optical pumping power was necessary to achieve inversion.

This Letter describes the characteristics of stimulated emission from trivalent uranium ions substituted for divalent calcium ions in calcium fluoride. The crystals were grown by a technique similar to that described by Galkin and Feofilov.[4] In contrast to the case of the ruby optical maser, the oscillation takes place in a transition from a metastable state to a level approximately 515 cm^{-1} above the ground state. At low temperatures this state is depopulated by at least a factor of 10^{10} relative to the ground state. The general condition for stimulated emission is that the number of excited ion in the metastable state must exceed the number of ions

* *Phys. Rev. Letters* **5**, 557 (1960).

in the lower state by a certain amount Δn sufficient to overcome the cavity losses. In trivalent uranium this condition will be achieved at low temperatures when the population of the metastable state is approximately equal to Δn since the population of the lower state is negligible. The value Δn may be generally much less than the number of ions in the ground state. It was indeed observed that the pumping power necessary to achieve stimulated emission was lower by about a factor of 500 compared to ruby.

Figure 1 shows the pertinent energy levels of the trivalent uranium ion when substituted for calcium in the CaF_2 lattice.[4] The

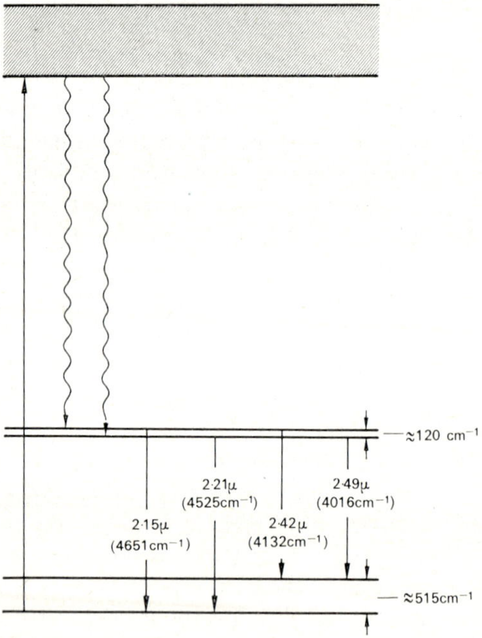

FIG. 1. Energy level diagram of trivalent uranium in calcium fluoride based on the paper by Galkin and Feofilov. Broad-band pumping light applied in the blue and green visible spectrum causes transitions to excited bands, one of which is depicted in the figure. These pumping transitions are followed by rapid, non-radiative transitions to the two metastable levels shown.

ion has several strong absorption bands in the visible and near infrared wavelength regions. The strongest absorption bands are in the green and blue, and as a result the crystals have a pale to deep red color and thus show a striking resemblance to ruby. Very strong fluorescence is observed in the infrared region between 2.0 and 2.6 microns when exciting radiation is applied in the range of the absorption bands. The room temperature fluorescence spectrum consists of four peaks. The two shorter wavelength peaks coincide with the absorption peaks and therefore are resonance lines. The two longer wavelength peaks represent emissions to a state located approximately 515 cm^{-1} above the ground state. The longest wavelength emission occurs from the lower metastable state and is the transition in which the maser oscillations are observed.

A single crystal of calcium fluoride with 0.05 mole percent uranium was cut into a cylindrical shape 3/8 in. in diameter and $1\frac{1}{2}$ in. long. The ends were polished flat to one-twentieth of a sodium wavelength and parallel to fifteen seconds of arc. Silver films were evaporated on both ends, with one end left completely opaque and the other end left with a 1.5% transmission. A protective SiO coating was applied over the silver to prevent oxidation. The sample was placed in a specially constructed liquid helium Dewar and was cooled without being in actual contact with the liquid helium. A pulsed xenon discharge lamp excited the vertically positioned sample through a side port in the Dewar. The infrared radiation generated in the sample was coupled out through the 1.5% transmitting end face and was detected by a lead sulfide detector through a top port. Either a 2.5-micron interference filter or a Perkin-Elmer monochromator was used to isolate the infrared radiation from the lamp background. The response time of the detector was approximately 40 microseconds for decaying radiation.

Figure 2 shows typical oscilloscope traces of the detector response to the intense monochromatic stimulated emission when the pump power is set above the threshold value. The signal during the first 450 microseconds of Fig. 2(a) is due primarily to

(a)

(b)

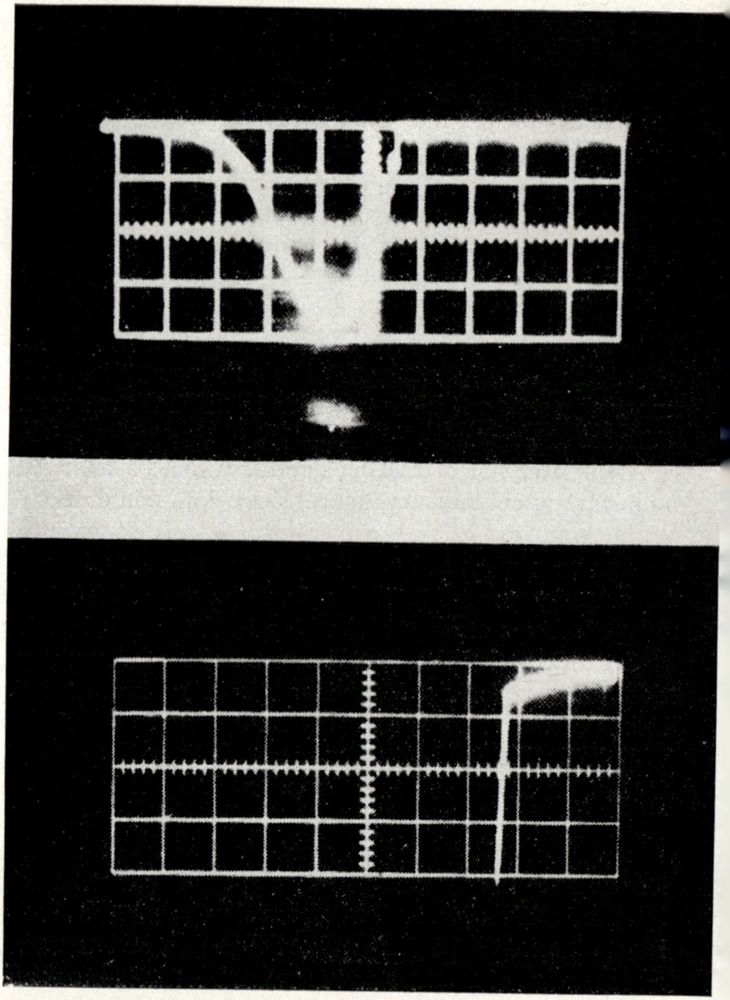

FIG. 2. (a) Detector response above threshold showing the onset of oscillation after 450 μsec. A filter with a bandwidth of 0.1 micron and centered at 2.5 microns is inserted before the detector. The sweep speed is 200 μsec/division. (b) Detector response with the conditions as in 2(a) except that the oscilloscope gain is reduced by a factor of 500 and the sweep speed is 100 μsec/division.

the scattered light from the xenon flash lamp plus a small but undetermined amount of spontaneous radiation. Oscillation begins where the rapidly rising, almost vertical, trace is observed and always occurs when the pumping power reaches the same critical level. Fig. 2(b) shows the same detector output with reduced gain and expanded time scale. The ratio between the stimulated emission and the background light was measured to be in excess of 1000, which means that the spontaneous emission is less than the stimulated emission by a factor of several thousand. This ratio is larger than that observed for the ruby maser, which conforms to the expectation of smaller spontaneous emission in the infrared as compared with the visible spectral region. The spectral width of the maser pulse was substantially less than the resolving power of the monochromator. The beam was confined to an angle of approximately 0.01 radian.

An interesting feature of the maser output shown in Fig. 2(b) is the presence of relaxation oscillations, previously also observed in less detail by Collins *et al.* in ruby. The number of observed spikes decreases rapidly as the power input into the exciting xenon lamp is lowered towards the threshold value until at barely above threshold only a single spike oscillation occurs. Figure 3 illustrates this phenomenon. The single spike takes place at exactly the peak of the pulse output of the xenon lamp, while the multiple spikes begin earlier because the exciting power reaches the threshold value at an earlier point. The spacing between the oscillations increases as the xenon lamp input power is brought closer to threshold. These features seem consistent with the explanation offered by Collins *et al.* for the relaxation oscillations in the ruby optical maser, perhaps complicated by contributions from several frequency modes within the inhomogeneously broadened line as suggested by G. J. Lasher of this laboratory.

The favorable energy level structure of trivalent uranium as compared to ruby may result in a strong cw infrared source for research purposes in the near future.

We would like to thank J. Lankard for assistance with the experiments, W. V. Smith and A. H. Nethercot for their continued

(a)

(b)

(c)

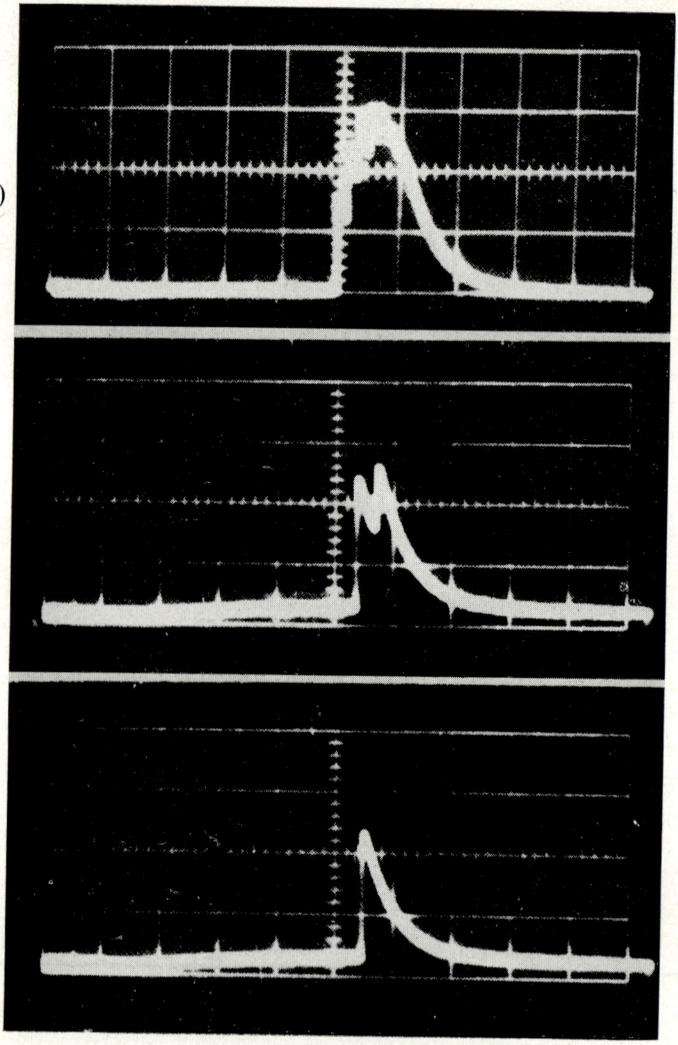

FIG. 3. Relaxation oscillations at varying pump power levels. The time scale is 100 μsec/division. Trace (a) shows the master output with pump power about 1.5% above threshold. Traces (b) and (c) are taken at powers barely above threshold.

interest and encouragement, and other members of the laboratory for helpful discussions. Special thanks are due to Walter Hargreaves of Optovac, Inc. for co-operating with us in growing the crystals.

References

1. A. L. Schawlow and C. H. Townes, *Phys. Rev.* **112,** 1940 (1958).
2. T. H. Maiman, *Nature* **187,** 493 (1960); *British Communications and Electronics* **7**, 674 (1960).
3. R. J. Collins, D. F. Nelson, A. L. Schawlow, W. Bond, C. G. B Garrett and W. Kaiser, *Phys. Rev. Letters* **5,** 303 (1960).
4. L. N. Galkin and P. P. Feofilov, *Doklady Akad. Nauk S.S.S.R.* **114,** 745 (1957) [translation: *Soviet Phys. — Doklady* **2,** 255 (1957)].

7

Stimulated Optical Emission in Fluorescent Solids. I. Theoretical Considerations*

T. H. MAIMAN†

Hughes Research Laboratories, A Division of Hughes Aircraft Company, Malibu, California

Summary

An analysis of stimulated emission processes in fluorescent solids is presented. The kinetic equations are discussed and expressions for pumping power and effective temperature of the exciting source are given in terms of the material parameters. A comparison of excitation intensity for three- and four-level systems is given. The spectral width of the stimulated radiation is discussed with particular attention to imperfect crystals.

Introduction

Experiments culminating in the achievment of stimulated optical radiation from Cr^{3+} in Al_2O_3 (ruby) were recently reported by the author[1,2]; in addition, stimulated emission from other solids has since been reported.[3] A discussion of stimulated emission processes in the infrared-optical spectral region was presented

Phys. Rev.* **123, 1145 (1961).

† Now at Quantatron, Inc., 2520 Colorado Avenue, Santa Monica, California.

[1] T. H. Maiman, *Phys. Rev. Letters* **4,** 564 (1960).

[2] T. H. Maiman, *Nature* **187,** 493 (1960): *Brit. Comm. and Electronics* **7,** 674 (1960); Program of the 45th Annual Meeting of the Optical Society of America, October 12–14, 1960 (unpublished), p. 14.

[3] P. P. Sorokin and M. J. Stevenson, *Phys. Rev. Letters* **5,** 557 (1960); *IBM J. Research Develop.* **5** (1961).

previously by Schawlow and Townes,[4] with particular emphasis on an alkali vapor system; however, the modifications of both the analytical and experimental problems encountered with solids are considerable. Some of these extensions and revisions are presented here.

The general class of materials to be considered are fluorescent solids whose emission spectra consist of one or more sharp spectral lines. Excitation is normally supplied to these solids by radiation of frequencies which produce absorption into one or more bands. Some of this excitation energy is lost by a combination of spontaneous emission and thermal relaxation to lower-lying states; however, if the solid has a relatively high fluorescent efficiency[1] most of the energy is transferred to the sharp fluorescent levels by means of a non-radiative process. Subsequently, by a combination of spontaneous emission and thermal relaxation, the excited atoms (ions) return either to the ground state, or another low-lying state. The spontaneous emission from these sharp levels is the observed fluorescent radiation. If the exciting radiation is sufficiently intense it is possible to obtain a population density in one of the fluorescent levels greater than that of the lower-lying terminal state. In this situation, spontaneously emitted (fluorescent) photons traveling through the crystal stimulate upper state atoms to radiate, and a net component of induced emission is superimposed on the spontaneous emission.

We distinguish at this point between two useful solid-state systems: a three-level scheme typified by the level diagram shown in Fig. 1, and a four-level arrangement as shown in Fig. 2. Other configurations are of course, possible, such as one in which all of the levels are relatively sharp and therefore more analogous to the gaseous situation. However, one of the features of the solid system is the possible use of a broad absorption band for the pump transition. This situation allows a relatively high pumping efficiency to be realized since most high-power optical sources have very broad spectral distributions in their radiant energy.

[4] A. L. Schawlow and C. H. Townes, *Phys. Rev.* **112,** 1940 (1958).

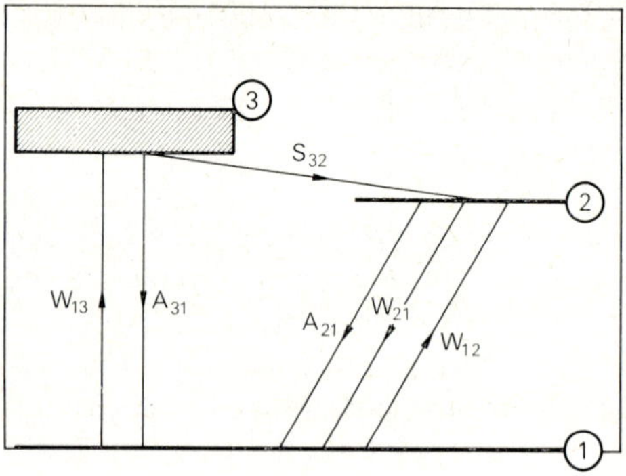

FIG. 1. Optical energy-level diagram for a three-level fluorescent solid.

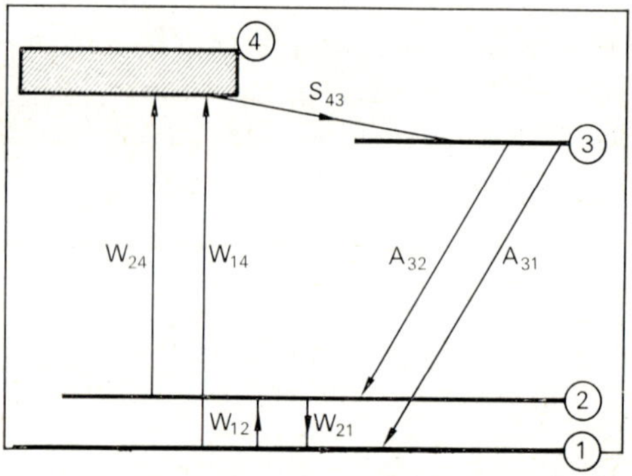

FIG. 2. Energy-level diagram in a fluorescent solid to illustrate a four-level method of producing stimulated optical radiation.

Pumping Power Considerations

Three-level System

In the configuration shown in Fig. 1, the ground state is the terminal level for the spontaneous emission transition $2 \rightarrow 1$. Therefore, to produce a net stimulated emission component it is necessary to have an incident pump radiation intensity large enough to excite at least one-half of the total number of ground-state atoms into level 2. The extreme difficulty of achieving such a radiation intensity led early investigators to ignore the three-level system here described.[5]

A more quantitative description of this scheme can be obtained from a solution of the steady-state rate equations

$$dN_3/dt = W_{13}N_1 - (W_{31} + A_{31} + S_{32})N_3 = 0, \tag{1}$$

$$dN_2/dt = W_{12}N_1 - (A_{21} + W_{21})N_2 + S_{32}N_3 = 0, \tag{2}$$

$$N_1 + N_2 + N_3 = N_0. \tag{3}$$

These processes are indicated in Fig. 1, where W_{13} is the induced ransition probability per unit time for the transition ($1 \rightarrow 3$) caused by the exciting radiation of frequency ν_{13}, W_{21} is the induced probability ($2 \rightarrow 1$) due to the presence of radiation of frequency ν_{21}, A_{31} and A_{21} are Einstein A coefficients to account for the spontaneous radiation, and S_{32} is the transition probability for the nonradiative process ($3 \rightarrow 2$). N_1, N_2, and N_3 are the respective level population densities, and N_0 is the total active ion density in the crystal, It has been assumed that the thermal processes for both ($2 \rightarrow 1$) and ($3 \rightarrow 1$) are negligible compared to the corresponding radiative processes but that the reverse is true for the transition ($3 \rightarrow 2$).[1] It is also assumed that each of the energy-level separations is large compared to kT so that the Boltzmann factors are negligible.

A solution to the above equations is

$$\frac{N_2}{N_1} = \left[W_{13}\left(\frac{S_{32}}{W_{31} + A_{31} + S_{32}}\right) + W_{12}\right] \Big/ (A_{21} + W_{21}). \tag{4}$$

If the solid has a high fluorescent quantum efficiency, e.g., ruby, $A_{31} \ll S_{32}$. Also, it will be found that even for very high pumping powers $W_{31} \ll S_{32}$. These approximations imply $N_3 \ll N_1$,

[5] A. L. Schawlow, in *Quantum Electronics*, edited by C. H. Townes (Columbia University Press, New York, 1960), pp. 553–563.

$N_3 \ll N_2$ and (4) simplifies to

$$N_2/N_1 \cong (W_{13} + W_{12})/(A_{21} + W_{21}), \qquad (5)$$

alternatively,

$$(N_2 - N_1)/N_0 \cong (W_{13} - A_{21})/(W_{13} + A_{21} + 2W_{12}). \qquad (6)$$

In order to obtain stimulated emission at the frequency ν_{12}, it is necessary that $N_2 > N_1$ and therefore, from (6) we must have $W_{13} > A_{21}$. This is a minimum condition. In order to achieve a useful stimulated-emission amplifier or oscillator, it is further required that the excess population $(N_2 - N_1)$ be sufficient to overcome circuit losses.[4] The degree of population inversion needed can be conveniently represented for the optical case as follows.

The energy in a wave propagating through a material with an absorption coefficient α is attenuated by the factor $e^{-\alpha l}$, if l is the length of material. The absorption coefficient is related to the population difference by

$$\alpha_{12} = (N_1 - N_2)\sigma_{12}, \qquad (7)$$

where σ_{12} is the absorption cross section for the pertinent transition. If $N_2 > N_1$, α is negative and $e^{-\alpha l} > 1$, i.e., amplification instead of absorption ensues. However, if $|\alpha|$ is small, a useful degree of amplification will be obtained only with an impractically long crystal. In such cases the interaction of radiation with the material can be increased by placing it between reflecting plates and advantage taken of the resulting regeneration due to waves reflected back and forth many times through the crystal. A wave that has traveled through the crystal once and reflected once has changed in energy content by the factor $e^{-\alpha l}r$, where r is the reflection coefficient of the end plate. In order to achieve a high amplification, or in the limit a coherent oscillation, the requirement is $e^{-\alpha l}r \cong 1$. In most cases $-\alpha l \ll 1$ so that the oscillation condition reduces to

$$-\alpha l \cong 1 - r \qquad (8)$$

or

$$(N_2 - N_1) = (1 - r)/\sigma_{12}l. \tag{9}$$

A different approach is to recognize that the parallel plate material system is an optical cavity, although not quite the same as a Fabry–Perot etalon because of the presence of the dielectric material; this latter point is discussed under "Spectral Width". The "Q" of the axial modes in this cavity can be obtained by solving for the mean decay time τ of photons moving perpendicular to the end plates.[4] When $(1 - r) \ll 1$,

$$\tau \cong l\varepsilon^{\frac{1}{2}}/[c(1 - r)], \tag{10}$$

where l is the distance between the plates and ε is the dielectric constant of the material between the plates. Therefore,

$$Q_c = \omega\tau = 2\pi l\varepsilon^{\frac{1}{2}}/[\lambda(1 - r)], \tag{11}$$

To account for the absorption (or emission) of energy within the material itself, we can define a material Q and find by similar reasoning to that above

$$Q_M = 2\pi\varepsilon^{\frac{1}{2}}/\alpha\lambda. \tag{12}$$

In these terms the oscillation condition is $Q_c \cong -Q_M$ and is seen to be the same as Eq. (8). If we denote as α_0 the normally measured absorption coefficient for the transition $1 \to 2$ under low-power excitation, then, at the point of oscillation,

$$(N_2 - N_1)/N_0 \cong (1 - r)/\alpha_0 l \tag{13}$$

for a typical situation $(1 - r) \ll \alpha_0 l$; therefore,

$$N_2/N_1 \cong 1 + 2(1 - r)/\alpha_0 l. \tag{14}$$

and only a small excess over equal populations is required.

As the pumping power is increased above the point where $W_{13} = A_{21}$ (equal populations), the crystal begins to emit stimulated as well as spontaneous radiation. The spontaneous emission power is $N_2 h\nu_{12} A_{21}$, and since N_2 normally increases by only a small amount over the equal population value, as indicated above, almost all additional power developed at the frequency ν_{21} appears as stimulated emission. Since only this latter component constitutes useful output it would be desirable to pump at levels where $W_{13} \gg A_{21}$; unfortunately, however, it is difficult in practice even to reach the condition $W_{13} = A_{21}$.

Consider a parallel beam of light incident on the surface of an optically thin sample of the material. The power in the beam is

$$P_0 = nh\nu_p, \tag{15}$$

where n is the number of incident photons per second in the spectral region of the absorption band and ν_p is the center frequency of the pumping light. From the definition of transition probability

$$W_{13} = n\sigma_p/A \tag{16}$$

where σ_p is the absorption cross section for the pumping transition $(1 \rightarrow 3)$ and A is the area of the crystal perpendicular to the incident beam. Therefore, the flux density to produce equal populations $(W_{13} = A_{21})$ is

$$F = P_0/A = A_{21}(h\nu_p/\sigma_p). \tag{17}$$

Thus, in the case of an optically thin sample $(\alpha_p l \ll 1)$ the required flux density is independent of crystal concentration. Of course, in this case, only a small fraction of the incident power is absorbed, but increasing the ion density in the crystal results in nonuniform pumping so that in practice a compromise is made. The problem is alleviated somewhat if the back surface of the crystal is coated with reflecting material or if the entire crystal surface is illuminated uniformly.

The energy density U_p in the crystal is related to the flux density by $U_p = F\varepsilon^{\frac{1}{2}}/c$ in the case of a parallel beam; therefore, using (17),

$$U_p = W_{13}h\varepsilon^{\frac{1}{2}}/\lambda_p\sigma_p. \tag{18}$$

Although derived for the case of a parallel beam, the energy density is the fundamental quantity that determines transition probabilities, and, therefore, (18) is correct for other pumping geometries as long as the crystal is optically thin. The flux density, on the other hand, will depend on the particular configuration used, [e.g., $F = (c/4\varepsilon^{\frac{1}{2}})U_p$ in the case of uniform illumination].

A configuration more likely to be realized in practice is one in which the material is more or less uniformly illuminated over most of its surface. Such an arrangement might be comprised of an appropriate reflector, lens, or reflector–lens combination. An alternative technique would be to use an extended source and place the active material in close proximity to it. For example, the source may be in the form of a hollow cylinder or helix with the active material placed inside. A disadvantage of the latter arrangement is that unused radiation from the exciting source aggravates heat dissipation problems; however, the over-all structure is much more compact.

The type of source most likely to be used is a high-pressure gaseous or vapor arc lamp. The spectral lines characteristic of the particular gas or vapor are practically nonexistent in such a lamp because of a combination of pressure and Stark broadening. The resultant output has some vague peaks, but is otherwise a more or less continuous spectrum with an effective radiation temperature typically between 4000°K and 10,000°K. Therefore, it is convenient and not too unrealistic, to consider that the active material is immersed in isotropic blackbody radiation.

We again assume an optically thin crystal so that the radiation density is uniform throughout its volume. Equation (18) is applicable to monochromatic radiation; in the case of broad-band radiation,

$$W_{13} = \int \frac{c\sigma(v)\rho(v)\,dv}{hv_p\varepsilon^{\frac{1}{2}}}, \tag{19}$$

where $\rho(v)$ is the energy density per unit volume per unit frequency interval. Taking $\sigma(v)$ as Lorentzian, and using

$$\rho(v) = \frac{8\pi hv^3\varepsilon^{\frac{3}{2}}}{c^3(e^{hv/kT} - 1)}, \tag{20}$$

we get

$$W_{13} \cong \frac{4\pi^2\varepsilon\,\Delta v_p\sigma_p v_p{}^2}{c^2(e^{hv/kT} - 1)}, \tag{21}$$

where $\Delta\nu_p$ is the half-width of the absorption band. From radiation theory,

$$A_{31} = 4\pi^2 \varepsilon \Delta\nu_p \sigma_p \nu_p{}^2/c^2 \tag{22}$$

for a Lorentzian line; therefore,

$$W_{31} = A_{31}/(e^{h\nu_p/kT_s} - 1 . \tag{23}$$

The source temperature required for equal populations ($N_2 = N_1$, $W_{13} = A_{21}$) is therefore

$$T_s = \frac{h\nu_p}{k \ln [1 + (A_{31}/A_{21})]} . \tag{24}$$

If the lifetime of the fluorescent level is not wholly radiative, $1/A_{21}$ should be replaced by τ_2, the actual lifetime of this state.

The important material parameters connected with pumping power requirements for the three-level system are indicated in (24). The effective source temperature is a measure of the power per unit area per unit frequency interval and is, therefore, germane to the pumping problem. Since the necessary temperature is proportional to the pump frequency ν_p, operation at shorter wavelengths becomes increasingly difficult. The main consideration at a given wavelength is to maximize the ratio $A_{31}:A_{21}$.

One of the main problems in obtaining stimulated emission from the three-level solid is the very high power density required for the pump source with its associated heat-dissipation problem. A possible solution is to use a pulsed rather than a continuous source, i.e., a high-power electronic flash lamp. If the pulse is short compared to the life of the fluorescent level ($\tau_s \ll \tau_2$) the atoms excited to the fluorescent level do not appreciably decay during such an exciting pulse. The total energy absorbed when equal populations are attained is $E = \frac{1}{2} N_0 h\nu_p V$. If the crystal is isotropically irradiated, the absorbed energy is also (from the definition of Q_M at the pump frequency)

$$E = (2\pi\nu_p U_p V/Q_M)\tau_s . \tag{25}$$

Combining with previously derived relations, we find that the required energy per unit area radiated in one pulse from the flash tube is

$$J = F\tau_s \cong h\nu_p/4\sigma_p. \tag{26}$$

If, on the other hand, the pulse is long compared to the fluorescent decay time then the necessary *peak power* developed by the source is the same as that derived previously for a continuous source. The *average* power, however, can be made arbitrarily small by keeping the flash rate low. If large average powers are desired, larger flash tubes and consequently larger crystals can be employed. Although it requires about the same average input power to obtain a given average output power, the power dissipation problem is much less severe with pulsed operation than with continuous operation because of the greater volume of material involved.

Four-level System

The level scheme depicted in Fig. 2 requires, in principle, lower exciting intensity than the three-level system discussed above.[5] If the low-lying level 2 is high enough above the ground state so that $e^{-h\nu_{12}/kT} \ll 1$, then level 2 will be relatively unpopulated. It will be as difficult as before to obtain an excess population with respect to the ground state, but negative temperatures for the transition $3 \rightarrow 2$ will be obtained at much lower intensities. The rate equations for this system are:

$$dN_3/dt = W_{14}N_1 + W_{24}N_2 - (A_{32} + A_{31})N_3 = 0, \tag{27}$$

$$dN_2/dt = w_{12}N_1 - (W_{24} + w_{21})N_2 + A_{32}N_3 = 0, \tag{28}$$

$$N_1 + N_2 + N_3 = N_0. \tag{29}$$

We have neglected the induced rates $W_{32} = W_{23}$ which do not become appreciable until inversion is produced and have assumed as previously that the nonradiative process S_{43} is much greater than either A_{42}, A_{41}, W_{14}, or W_{24}. The solution is then

$$\frac{N_3}{N_2} \cong \frac{W_{14}(W_{24} + w_{21}) + (W_{24} + w_{12})W_{24}}{W_{14}A_{32} + w_{12}(A_{32} + A_{31})}. \tag{30}$$

The quantities w_{12} and w_{21} are thermal transition probabilities, and the other quantities have the same meaning as previously stated. Since at room temperature thermal relaxation between levels 1 and 2 is apt to be very fast ($\sim 10^{-6}$ sec), $w_{12} \gg W_{14}$ and $w_{21} \gg W_{24}$; also the thermal probabilities are related by $w_{12} = w_{21} \exp[-h\nu_{12}/kT]$. Therefore,

$$\left(\frac{N_3}{N_2}\right) \cong \frac{W_{14}e^{h\nu_{12}/kT} + W_{24}}{A_{32} + A_{31}} \tag{31}$$

and

$$N_2/N_1 \cong e^{-h\nu_{12}/kT}.$$

If we make the further assumption $W_{14} \approx W_{24}$, and define $A_3 = A_{31} + A_{32}$, we find that

$$\frac{N_3 - N_2}{N_0} \cong \frac{W_{14} - A_3/(1 + e^{h\nu_{12}/kT})}{W_{14} + A_3}. \tag{32}$$

In order to obtain high gain or oscillation the excess population must be

$$N_3 - N_2 \cong (1 - r)/\sigma_{32}l. \tag{33}$$

Combining (32) and (33) and assuming $N_0\sigma_{32}l \gg 1 - r$, we find

$$W_{14}\tau \cong \frac{1 - r}{N_0\sigma_{32}l} + \frac{1}{1 + e^{h\nu_{12}/kT}}, \tag{34}$$

where $\tau = 1/A_3$ is the lifetime of the fluorescent level. This result is to be compared with the equivalent requirement $W\tau \approx 1$ in the three-level system.

Spectral Width

The spectral width of the radiation emitted from an optical maser is discussed by Schawlow and Townes in the ideal limiting case in which the oscillations excite only a single mode in a Fabry-Perot which the oscillations excite only a single mode in a Fabry–Perot type of resonator. This situation is possible with extremely precise fabrications in a gaseous or vapor system but has not been realized in solid state devices.[2,3,6] The effect of the solid dielectric is twofold. The first is the introduction of "light pipe" modes of propagation, due to internal reflection; the second is distortion of the mode patterns because of strains, inhomogeneities, and deviations from single crystallinity. The first effect is not so serious since it can, in principle, be circumvented, for example, by using a roughened cylindrical surface or immersing the maser material in a liquid (or solid) having nearly the same or higher refractive index; however, this is an additional problem to deal with.

Temperature effects in a solid are apt to be more serious than in a gaseous system. Generally speaking solids require a much higher power density because of the much larger density of atoms. For this reason, the magnitude of temperature shifts will be larger, and in addition the same temperature shifts will usually be of more consequence. The chamber for a gaseous system can be made of quartz or other special low temperature material whereas the plate separation in the solid situation is determined by the maser material itself which in turn may have a relatively high expansion coefficient. Finally, the energy levels are temperature dependent in a solid [about 3 (kMc/sec)/°C near room temperature for the ruby R_1 line].

Crystal imperfections, light-pipe modes, and endplate misalignments generally introduce ambiguity in the mode selection process; i.e., there may be a very large number of modes which are of equal status in contrast to the perfect situation where the axial (or nearly axial) cavity mode closest to the atomic resonance peak is clearly the preferred one. Temperature shifts give rise to frequency sweeping and mode hopping.

As discussed by Schawlow and Townes,[4] multimoding introduces additional noise into the system due to spontaneous emission considerations; in addition, as shown in any text on oscillator theory, this same noise input reduces the monochromaticity of the output radiation from the system when it is oscillating.

We consider a system such as described above in which a large

[6] R. J. Collins, D. F. Nelson, A. L. Schawlow, W. Bond, C. G. B. Garrett and W. Kaiser, *Phys. Rev. Letters* **5**, 303 (1960).

number of cavity modes enter into the oscillation because of poor mode selection. For purposes of illustration, we consider a cylindrical rod of three-level fluorescent material (such as ruby) with reflecting end plates to form an optical resonator. The power dissipated in this cavity is the sum of spontaneous and stimulated emission components. The stimulated emission power results primarily from the energy density built up in the cavity by standing waves and so essentially all of the stimulated emission power radiated by the material is coupled to the cavity. On the other hand, only a fraction (denoted by f) of the spontaneous emission power is dissipated in and excites the cavity. When regeneration ensues due to excess upper-state population, only the highest Q (axial and near axial) modes are excited appreciably. Although in practice only a small fraction of the total number of modes may be excited (limited principally by crystal perfection), the absolute number might be very large. (In a ruby of one-cm dimensions, the total number of modes is $\sim 2 \times 10^{11}$.)

The power per unit volume coupled to the cavity is

$$\begin{aligned} p_c/V &= 2\pi\nu_{12}U/Q_c \\ &= fA_{21}N_2h\nu_{12} + W_{21}(N_2 - N_1)h\nu_{12}\,. \end{aligned} \tag{35}$$

Here U is the energy density in the cavity, and Q_c is the loaded cavity Q (not including the material but including output coupling). Because we are dealing with a multimode cavity with dimensions very large compared to a wavelength, the spectral distribution of the spontaneous radiation is essentially the same as the fluorescence radiating into free space. If we assume a Lorentzian line, the frequency distribution of spontaneous power is $N_2A_{21}h\nu g(\nu)$, where

$$g(\nu) = \frac{\Delta\nu}{2\pi}\frac{1}{(\nu - \nu_0)^2 + (\Delta\nu/2)^2} \tag{36}$$

and $\Delta\nu$ is the half-power width of the atomic transition $(2 \rightarrow 1)$. Analogous to Eq. (19), we have

$$W_{21} = \int \frac{c\sigma(\nu)\rho(\nu)\,d\nu}{h\nu\varepsilon^{\frac{1}{2}}}, \tag{37}$$

where $\sigma(\nu) = \frac{1}{2}\pi\Delta\nu\sigma_0 g(\nu)$. The power per unit volume per unit frequency interval is then

$$\frac{2\pi\nu\rho(\nu)}{Q_c} = fN_2A_{21}h\nu g(\nu) + \frac{c}{\varepsilon^{\frac{1}{2}}}\sigma(\nu)\rho(\nu)(N_2 - N_1), \tag{38}$$

therefore, using (12)

$$\rho(\nu) = \frac{1}{4\pi^2}\frac{fN_2A_{21}h\Delta\nu Q_c}{(\nu-\nu_0)^2 + (\Delta\nu/2)^2(1 + Q_c/Q_M)}. \tag{39}$$

Thus, the energy density is Lorentzian with a half-amplitude width

$$B = \Delta\nu(1 + Q_c/Q_M)^{\frac{1}{2}}. \tag{40}$$

The total power delivered to the cavity is

$$P_c = \frac{2\pi\nu V}{Q_c}\int\rho(\nu)\,d\nu,$$

therefore, using (36) and (37),

$$B = fN_2A_{21}h\nu\Delta\nu V/P_c. \tag{41}$$

The fraction f in terms of cavity modes is well known to be

$$f = \frac{\bar{n}}{4\pi^2\nu^2\Delta\nu V\varepsilon^{\frac{3}{2}}/c^3}, \tag{42}$$

where the denominator is the total number of modes within the integrated width ($\frac{1}{2}\pi\Delta\nu$) of the atomic transition and $\bar{n}$ is the actual number of modes participating in the oscillation; also since

$$A_{21} = 4\pi^2\varepsilon\Delta\nu\sigma\nu^2/c^2,$$

we find

$$B = \bar{n}cN_2\sigma h\nu\Delta\nu/\varepsilon^{\frac{1}{2}}P_c. \tag{43}$$

Then, using $N_2/N_1 = e^{-h\nu/kT_e}$ (where T_e is the effective temperature of the two-level system composed of the fluorescent level and the ground state), $|Q_m| = 2\pi\varepsilon^{\frac{1}{2}}/|\alpha|\lambda \cong Q_c$ (when the system is oscillating), and $Q_c = \nu/\delta\nu$ (where $\delta\nu$ is the width of a single cavity mode) we obtain from (43)

$$B = 2\pi\bar{n}h\nu\Delta\nu\delta\nu/P_c(1 - e^{h\nu/kT_e}) \tag{44}$$

(T_e is a negative quantity).

An alternative expression for the spectral bandwidth is obtained by considering the total power developed at the frequency ν_{12},

$$P_t = N_1 W_{13} h\nu_{12} V = P_c + (1 - f)N_2 A_{21} h\nu V, \tag{45}$$

and using (41), we find

$$B = \frac{f\,\Delta\nu}{(W_{13}/A_{21})(N_1/N_2) - (1 - f)}. \tag{46}$$

When the pumping power is increased to the point where $W_{13} = A_{21}$ ($N_1 = N_2$), from (46) $B = \Delta\nu$ independent of f. Any further increase in pumping power results in the production of stimulated emission and narrowing of the spectral width. Equation (46) is plotted in this region for various values of the parameter f and shown in Fig. 3. Thus, with a fairly good crystal, where good mode selection is feasible ($f \ll 1$), an abrupt change in the spectral width occurs with a fairly well-defined threshold pump power, but with a badly strained crystal ($f \sim 1$), the energy is scattered into many modes and no well-defined threshold can be found. Both types of behavior have been observed (see following paper).

The validity of the foregoing analysis breaks down when the spectral width of the "mode packet" is comparable to, or less than, that of the atomic resonance. In this case account must be taken of the modified spectral distribution of the spontaneous radiation as discussed by Purcell[7] and Pound.[8] In the extreme case of a single mode oscillation one can use a wire circuit analysis[8,9] and arrive at an equation analogous to (44), i.e.,

$$B = \frac{2\pi h\nu(\delta\nu)^2}{P_c(1 - e^{h\nu/kT_e})}. \tag{47}$$

[7] E. M. Purcell, *Phys. Rev.* **89,** 681(A) (1946).
[8] R. V. Pound, *Ann. Phys.* (*N. Y.*) **1,** 24 (1957).
[9] T. H. Maiman, in *Quantum Electronics*, edited by C. H. Townes (Columbia University Press, New York, 1960), pp. 324–332.

This expression is similar to the one presented by Schawlow and Townes,[4] but more general in that account is taken of a finite lower state population (their result implies $T_e \rightarrow -0$). Also the width of the cavity mode appears here, as it should, since usually (for both solids and gases) $\delta\nu < \Delta\nu$ and the narrower of these two is the dominant element in determining the spectral width.

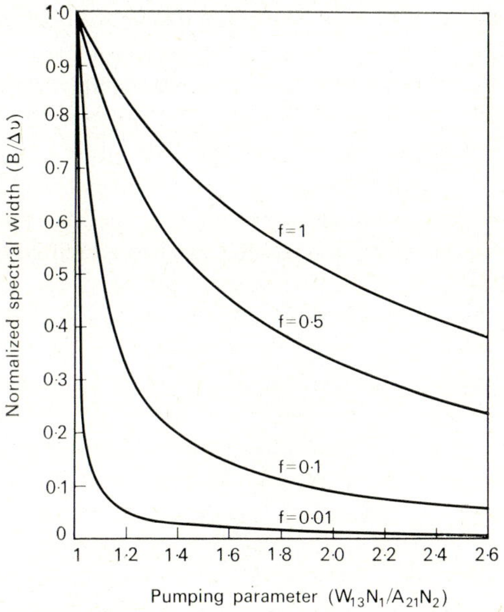

FIG. 3. Spectral bandwidth of the stimulated emission as a function of pumping power and the fractional number of cavity modes entering into the oscillation.

Conclusions

Optical excitation into a broad absorption band is possible with many fluorescent solids; thus, stimulated emission may be produced with relatively efficient utilization of typical broad band pumping sources. A three-level solid system as described here (e.g., ruby), uses the ground state as the terminal level for the stimulated emission transition. The required pumping power for this scheme is quite high; however, it has been shown that the important pumping parameter is the effective temperature of the source. A four-level fluorescent solid may also utilize a broad absorption band; however, in some cases, depending on the material parameters, much less pumping power is required than for the three-level scheme.

Solid-state systems pose additional problems in mode selection not present in a gaseous system. A dielectric rod consitutes a waveguide (light pipe) and thus additional modes of propagation (which may be low loss) are introduced. Even more important are strains, inhomogeneities, and deviations from single crystallinity which tend to scatter energy into undesired modes. Under the assumption that the mode selection is poor, i.e., many modes participate in the oscillation, an analysis was presented for the case of badly strained crystals. The general features of this formulation have been observed in the laboratory as discussed in the following paper.

No attempt has been made here to treat the transient behavior or multiple spectrum observed in pulsed excitation of ruby and other solids. A quantitative description would require an analytical expression for the variation of Q of the various mode possibilities. This in turn is a function of the particular crystal and its history of fabrication (see following paper).

8

Possibility of Production of Negative Temperature in Gas Discharges*

A. JAVAN

Bell Telephone Laboratories, Murray Hill, New Jersey

IN A recent paper[1] Schawlow and Townes have discussed a possibility for obtaining maser action in the optical region. In their proposed scheme, negative temperature is obtained by optical pumping. One may expect also that under favorable conditions the excitation of atomic levels by electrons in a discharge can lead, in principle, to a state of negative temperature. However, severe restrictions exist if densities of the excited atoms as large as those needed for maser action are required. The present Letter considers briefly these limitations and certain types of systems which appear to be most favorable for practical application of this proposal. Pure gases behave quite differently than certain kinds of gas mixtures. First let us consider the former case.

For the purposes of rough estimates of various discharge conditions as described below, let us make a simplifying assumption that the main source of population of an excited state is due to collisions of the first kind between the electrons and atoms in the ground state leading directly to the excited level under consideration. It can be shown that, at least in cases discussed below, other details such as cascade processes and collision of electrons with other excited atoms do not appreciably affect our order-of-magnitude estimates.

Phys. Rev. Letters* **3, 87 (1959).

In the following, the indices 0, 1, and 2 refer to the ground, the lower, and the upper states of the maser levels, respectively. If $1/\theta_{0i}$ is the rate of excitation of the ith excited state by electron impact with the atoms in the ground state and $1/\theta_{i0}$ is the rate for the reverse process, for a Maxwellian distribution of electrons at a temperature T_e, the relationship

$$\theta_{0i}/\theta_{i0} = \exp(E_i/kT_e)$$

will hold;[2] where E_i is the energy of the ith excited state. For θ_{i0} much shorter than the radiative lifetimes of the ith excited state, T_i, it can be shown that $N_i/N_0 = \exp(-E_i/kT_e)$, which means a Boltzmann equilibrium for population distribution of the excited state at the electron temperature T_e. For $\theta_{i0} \gg T_i$, however, one obtains

$$N_i/N_0 = (T_i/\theta_{i0}) \exp(-E_i/kT_e). \tag{1}$$

The ratio T_i/θ_{i0} is a measure of departure from the Boltzmann equilibrium. Such a departure is necessary for the problem under consideration. The factor $1/\theta_{i0}$ is proportional to the electron density in the discharge. It is therefore clear that excessively large electron densities should be avoided in order to satisfy $\theta_{i0} \gg T_i$, so that

$$N_2/N_1 = (T_2/T_1)(\theta_{10}/\theta_{20}) \exp[-(E_2 - E_1)/kT_e]$$

may be larger than unity, giving a negative temperature.

If the nature of the excited levels to be used for the maser action permits, it is very desirable to operate at a large gas pressure consistent with practical considerations. This is evident since for a given number of atoms in an excited state an effective temperature as determined from the ratio of this number to that of the atoms in the ground state is lower at larger pressure.

At a large pressure, however, the trapping of resonant radiation becomes an important factor. Consider a level which is optically connected to the ground state. At a pressure where the resonant photons arising from this transition are trapped, the effective radiative lifetime of this level will be increased and determined by its decay into the other available low-lying levels. This effect can best be taken to the advantage if the upper maser level is optically allowed to the ground state. In such a situation, in addition to an increased lifetime of the upper maser level at a large pressure, one expects also a more favorable ratio of θ_{10}/θ_{20}. This is evident since a level which is optically connected to the ground level, as a rule, has a large excitation cross section by electron impact with the atoms in the ground state. Furthermore, due to the nature of the electrical dipole matrix elements, only one of the maser levels may be optically connected to the ground level.

For a typical example of levels which allow operation at a large pressure, let us consider the case of Ne. The $2s_4$ and $2p_{10}$ (in Paschen notation)[3] considered as the upper and the lower maser levels, respectively, meet the required conditions. At a pressure where the ultraviolet photons arising from the transition $2s_4$ to the ground state are completely trapped, the lifetime of this state is estimated to be at least a factor of 60 larger than that of the $2p_{10}$ level. The latter level is known to have a mean life of 10^{-8} sec due to its radiative decay into some of the lower excited states. Let us consider 10 mm of Hg as a typical gas pressure. For production of $N_2 = 10^{10}/\text{cm}^3$, one obtains from Eq. (1), $T_2/\theta_{20} = 3 \times 10^{-5}$, assuming an electron temperature at $kT_e = 2.9$ ev. The cross section for electron excitation of the above Ne level is not known. Let us assume a reasonable value of $\sigma = 10^{-17}$ cm^2 for an average of this cross section over the electron energy distribution. From the above value of T_2/θ_{20} and $T_2 = 6 \times 10^{-7}$ sec, one obtains an electron density of $n_e = 5 \times 10^{10}/\text{cm}^3$.

The numbers obtained above are intended to represent only rough estimates and indicate that, in practice, the necessary discharge conditions are obtainable. Considerations such as departure from Maxwellian energy distribution of the electrons at the high- and low-energy limits may show appreciable effects. However, with a view to the possibility of operating at somewhat of a higher electron temperature, no serious consequences are expected.

In the above scheme, the presence of a large density of the metastable Ne should be avoided to prevent the trapping of the red Ne line which may lead to an extended lifetime for the $2p_{10}$ level.

Furthermore, the collision of the first kind of the electrons with the Ne metastables may give rise to an unfavorable population of the excited levels. This can be prevented by introducing a very small amount of a quenching gas such as argon to decrease the lifetimes of the longer lived excited atoms such as the metastables.

It may be of interest to compare the above system with a case which does not permit operation at large pressure. In He the 3^1D and 2^1P levels have a lifetime ratio of $T_2/T_1 = 35$. The lower level, 2^1P, has a short lifetime due to its radiative decay into the ground level. An estimate of θ_{10}/θ_{20} indicates that this ratio may be as large as 1/15. Thus, a negative temperature may still be expected if appreciable excitation of these levels by electron impact with the metastable He can be prevented. On the basis of the known cross section for electron excitation of the 3^1D level and manipulations similar to those presented above, the following rough estimate of the discharge condition may be obtained: At a pressure of 5×10^{-3} mm and for $kT_e = 5$ ev, the electron density for production of $N_2 = 10^9/\text{cm}^3$ is roughly $n_e = 10^{12}$ or about 1% ionization of the total number of atoms. Although such a condition is not outside of the range of practical possibility, however, it indicates the difficulties encountered in this case. It is likely that a more favorable situation of this type may be found in a different atom with a larger electron excitation cross section, thus resulting in a reduced number for the required electron density.

An alternative scheme is the use of transfer of excitation between excited states of two different atoms in a gas mixture. Consider a long-lived state of an atom (such as a metastable state). This state can be populated appreciably at moderate electron densities. If an excited state of a second atom happens to lie very close in energy to that of the level of the first atom, a large cross section is expected to exist for an inelastic collision resulting in a transfer of excitation from the metastable state to the excited state of the other atom and vice versa. Due to the nonadiabatic nature of the process of collision, the levels of the second atom which differ in energy considerably from that of the metastable level of the first atom do not show appreciable cross sections for transfer of excitation.

If the rate at which such transfer of excitation takes place is larger than the rate of the radiative decay of the excited states, the metastable level serves to increase the contact between the temperature of the excited level of the other atoms and that of the electrons. In this case, it can be shown that $(G^*N^*/GN) = (g^* n^*/gn)f$, where the capital letters refer to one atomic species and the small letters to the other. The letters with asterisks denote the excited states, and those without asterisks, the ground states; G and g are the statistical weights of the levels involved and N and n are the populations of the levels. The factor f in this equation is given by $f = e^{\Delta E/kT}$, where T is the temperature of the gas as determined by thermal kinetic energies of the atoms and ΔE is the resonance defect and is the difference between the excitation energy of the level corresponding to n^* minus that corresponding to N^*.

This relationship does not hold if ΔE is appreciably large, resulting in reduced rate of transfer of excitation.

In order to take full advantage of the above effect, large differences in the partial pressures of the two atoms are desirable. Otherwise, due to reduction in electron temperature and various quenching mechanisms, the population of the metastable level will be decreased.

As an example, a mixture of Kr and Hg, with Kr at about 10 to 50 mm of Hg pressure and Hg at about 10^{-3} mm of Hg pressure, can be used for obtaining an abnormally large population of the 9^1P and 6^1F of Hg. These two levels lie very close to $5s_5$ metastable Kr. A negative temperature at the transition originating from these levels to 6^1D may then be expected. It is expected that the levels 6^1D and 6^1P may also show a negative temperature because of the cascade transition from the 9^1P and 6^1F into 6^1D level.

The $^3S_1$ metastable of He also lies in energy fairly close to the excitation energy of the upper maser level of Ne discussed above. The presence of a partial pressure of He is expected to enhance considerably the negative temperature in the levels of Ne.

The transfer of excitation of the type described above may play an appreciable role within the levels of the same atom. An important example of this is expected to occur in the levels of Ne. Let us consider the group of four levels $2s_5$, $2s_4$, $2s_3$, and $2s_2$. The level $2s_4$ is the one emphasized in the above for the upper of the two maser levels. These four levels all fall fairly close in energy. The level $2s_2$ is also allowed for an optical transition to the ground state. The transfer of excitation within these levels is expected to result in particular build up of large population in the level $2s_5$, this level having the lowest energy within this group. Thus, an even more favorable transition in Ne appears to be the $2s_5 \rightarrow 2p_{10}$. This transition lies at 10343 wave number.

Details of the above proposals will be published upon their experimental verifications.

I would like to acknowledge helpful discussions with Dr. W. R. Bennett, Jr., Dr. S. J. Buchsbaum, Professor C. H. Townes, Dr. J. P. Gordon and Dr. A. L. Schawlow.

References

1. A. L. SCHAWLOW and C. H. TOWNES, *Phys. Rev.* **112,** 1940 (1958).
2. See, for instance, RUDOLF LADENBERG, *Revs. Modern Phys.* **5,** 243 (1933).
3. In the *LS* designations the levels $2s_4$ and $2p_{10}$ correspond to $(2p^5 4s)^3P_1$ and $(2p^5 3p)^3S_1$, respectively.

9

Population Inversion and Continuous Optical Maser Oscillation in a Gas Discharge Containing a He–Ne Mixture*

A. JAVAN, W. R. BENNETT, JR. AND D. R. HERRIOTT

Bell Telephone Laboratories, Murray Hill, New Jersey

THIS Letter is intended to give a summary of the results of experimental determinations of several physical properties of a gaseous discharge consisting of a He–Ne mixture which have led to the successful operation of a continuous-wave maser at five different wavelengths in the near-infrared. Population inversions are achieved between several Ne levels by means of excitation transfer from the metastable He($2\ ^3S$) to the $2s$ levels of Ne.[1,2] The maser oscillation takes place in a narrow beam with a diameter of 0.45 inch. The power in our strongest beam at the wavelength 11 530 Å is 15 milliwatts. The measured linewidth is in the range of 10 to 80 kc/sec. The angular spread of the beam is less than one minute of arc.

In the present system, the He($2\ ^3S$) metastables are used as carriers of energy to excite the levels of Ne. Due to the resonant nature of this process, the excitation cross section is large only to those levels of Ne which fall in energy within a few kT of that of the He($2\ ^3S$). Figure 1 gives the pertinent energy levels of He and Ne. As may be seen from Fig. 1,[3] the energy separations between the He($2\ ^3S$) and Ne($2s$) states are small enough to insure an appreciable degree of transfer to the $2s$ states; however, the

* *Phys. Rev. Letters* **6,** 106 (1961).

M

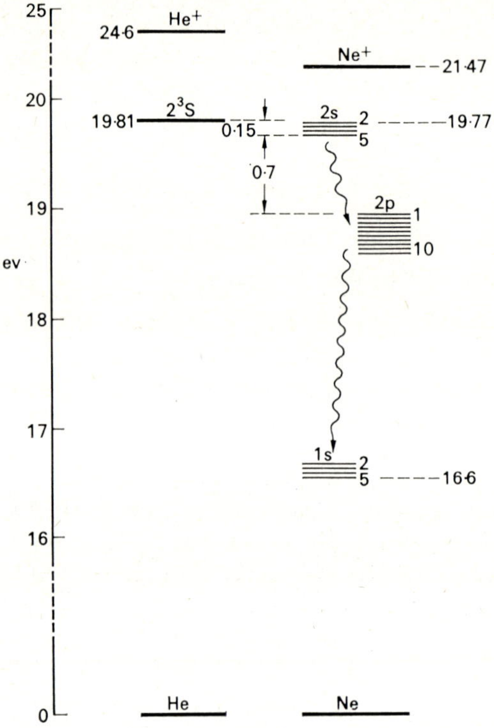

FIG. 1. Energy level diagram of He and Ne atoms.

energy gap between the 2 ^3S and $2p$ levels is much too large to permit direct population of the $2p$ states in the collision process. Although the $2s_2$ and $2s_4$ levels of Ne may radiate to the Ne ground state, in the limit of complete resonance trapping their lifetimes are determined primarily by radiative decay to the $2p$ levels. Under this condition, the lifetimes of all of the $2s$ levels are about one order of magnitude longer than those of the corresponding $2p$ states. (The decay of the $2p$ levels is due to their radiative decay to the $1s$ levels.) Thus population inversions may be obtained on each of the thirty allowed $2s \rightarrow 2p$ transitions (see later).

In order to determine the transfer cross section and spurious sources of excitation of Ne levels, the afterglow of a pulsed rf discharge containing varying amounts of He and Ne was studied in detail. The time variation of the densities of the $2s$ and $2p$ levels was observed by the fluorescent decay of these levels using appropriate isolated transitions. The time variation of the $2\,^3S$ density was determined through the absorption of the $2\,^3P \rightarrow 2\,^3S$ He transition.

The conditions were found under which the dominant mode of excitation of Ne levels arose from the He($2\,^3S$) transfer. Among other sources of excitation of the Ne levels it was found that inelastic collisions of energetic electrons with Ne metastables[2] (i.e., the $1s$ levels[4]) were by far the most important. It should be noted that certain amounts of $2p$ excitation, arising from sources other than $2s - 2p$ cascade, can be tolerated without upsetting the presence of a population inversion. For instance, let us consider the population inversion between the $2s_5$ and $2p_9$ levels. The $2p_9$ is optically connected to only one of the $2s$ levels, namely the $2s_5$. The Einstein A coefficient for the $2s_5 \rightarrow 2p_9$ transition was measured and found to be at least 25 times smaller than the rate of decay of the $2p_9$ level to within an accuracy of 20%. (This measurement was done by means of two independent techniques. The first was a spectroscopic technique to be described in a later publication. The second is discussed below.) This means that even an order of magnitude added excitation of $2p_9$ over that due to the cascade from the $2s_5$ still yields a population of the $2s_5$ levels larger than twice the population of the $2p_9$. In view of this effect, the adjustment of the electron density was found to be most critical.

In the afterglow, the energetic electrons thermalize within 10 to 20 μsec. The excitation transfer from He($2\,^3S$), however, continues until the time when all of the $2\,^3S$ states are quenched. This effect gives rise to a decay of the $2s$ levels identical to the time variation of the He($2\,^3S$). The afterglow of the $2p$ levels also shows an identical decay due to the cascade transition from the $2s$ levels. Our observations indicate that the time variations of the

$2\,^3S$, $2s$, $2p$ levels in the afterglow are indeed identical to a high degree of accuracy. The possibility of the direct transfer from the $2\,^3S$ to the $2p$ levels was ruled out by a process of relative intensity measurement at a fixed time in the afterglow to be described elsewhere.

Study of the fluorescent decay rate of the $2s$ and $2p$ levels as a function of several Ne pressures gave a total velocity-averaged inelastic cross section of $\langle\sigma\rangle = (3.7 \pm 0.5) \times 10^{-17}$ cm^2 for destroying He($2\,^3S$) metastables in two-body collisions with Ne ground-state atoms at 300°K. This cross section can safely be assigned to the excitation transfer to the $2s$ levels of Ne since we have been able to show that other sources of destruction of the He($2\,^3S$) by Ne ground-state atoms such as the one leading[5] to a formation of $HeNe^+$ are negligible. Our cross section measurement at the same time furnishes the value of the diffusion coefficient of $2\,^3S$ He metastables. Our result is in excellent agreement with the value of (470 ± 25 cm^2/sec) determined by Phelps.[6]

The radiative lifetimes of the $2s$ and $2p$ levels were determined using techniques described previously.[7] Our tentative values of the lifetimes of the $2s_2$, $2s_3$, $2s_4$, and $2s_5$ fall in the range from about 100 to 200 mμsec. Our results for the $2p$ levels are in general agreement with Ladenberg's[8] results. These measurements are still in progress and will be reported more fully at a later date. Each of the $2s$ levels decays into several of the ten $2p$ levels. The measured and estimated branching ratios for the transitions originating from each of the $2s$ levels, together with their total rate of radiative decay obtained from the above lifetimes, give the values of the individual Einstein A coefficients connecting the various $2s$ and $2p$ levels.

From the early stages of this experiment we were able to observe single-pass amplification of a transmitted signal at an infrared frequency where the inverted population was expected. Several electronic detection schemes were devised for this purpose. One of these schemes responded only during a small portion of the time and was in synchronism with the rf pulsed discharge. This technique enabled us to observe a larger value of amplification in the afterglow than during the glow; the latter is to be expected since the added excitation of the $2p$ levels by electron impact is absent in the afterglow. However, in addition to extreme difficulties encountered in the interpretation of the results, the gain measurement alone was not sufficient to shed insight into the nature of the physical processes involved in the discharge.

In view of the above considerations, we were able to determine an optimum condition for the maser action prior to the observation of oscillation as described below.

The continuous-wave maser oscillation is observed in a discharge containing 0.1 mm Hg of Ne and 1 mm Hg of He. The optical feedback[9] is obtained through high-reflectance Fabry-Perot plates placed within the gas chamber. This chamber consists of a long quartz tube with an inside diameter of 1.5 cm and length of 80 cm. The quartz tube is terminated at each end with a larger metal chamber containing the high-reflectance plates. Flexible bellows are used in the end chambers to allow external mechanical tuning of the Fabry-Perot plates. Two optically flat windows at the extremes of this system are provided to allow the maser beam to leave the instrument undistorted. The separation of the plates is one meter. The discharge is excited by means of external electrodes using a 28-Mc/sec generator. The estimated power dissipation is around 50 watts.

The high reflectance is achieved by means of 13-layer evaporated dielectric films on the hundredth wavelength flat surface of the fused silica plates. The reflectance is 98.9% with 0.3% transmission in the wavelength range 11 000 Å to 12 000 Å. Outside of this range of wavelength the reflectance decreases rapidly.

The initial adjustment of the plates prior to starting of the discharge is done using an autocollimator. We have observed oscillations at the following wavelengths: 11 180, 11 530, 11 600, 11 990, and 12 070 Å. The strongest oscillation occurs at 11 530 Å with an output power of 15 mw.

An image convertor tube is used visually and with a camera to observe and record structure of the beam. Two types of beam patterns have been studied.

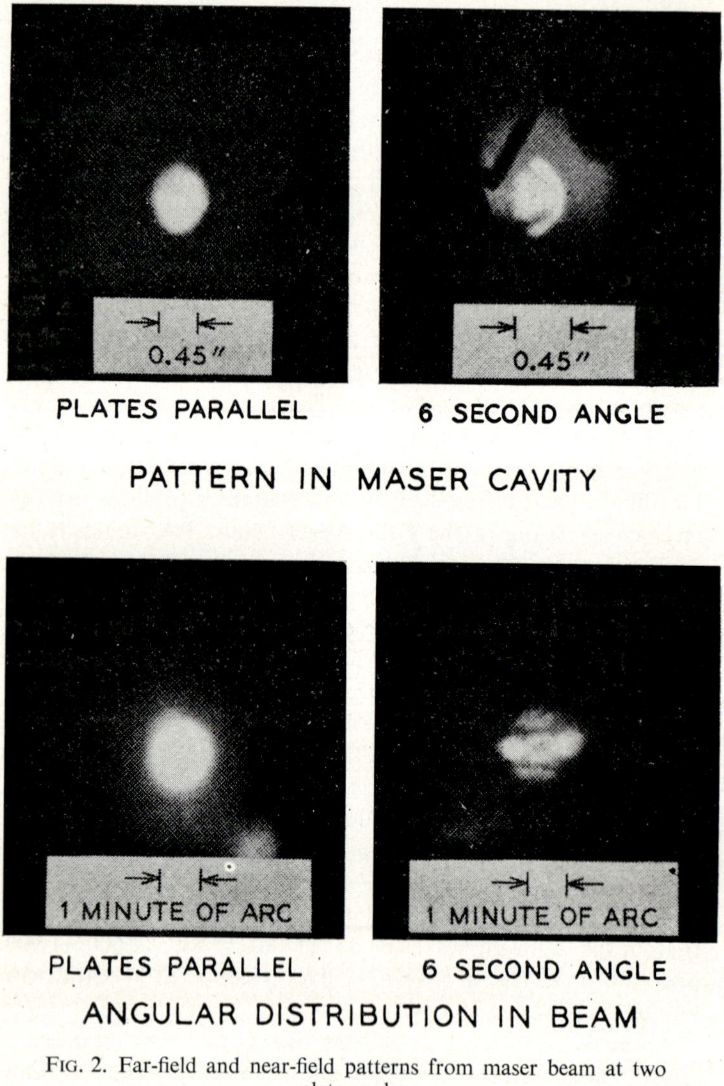

FIG. 2. Far-field and near-field patterns from maser beam at two plate angles.

A cross section of the emergent beam can be intercepted by the photocathode of the image convertor and the intensity distribution recorded. This near-field pattern is shown in the two upper photographs of Fig. 2 for two Fabry-Perot plate angles.

The intensity pattern that the beam would form at an infinite distance can be observed by imaging the maser beam to a point and observing the distribution of light in this image with an image convertor microscope. The intensity distribution in this image represents the angular distribution in the maser beam. The two lower pictures in Fig. 2 show this far-field pattern at the indicated plate positions. The structures seen in the near- and far-field patterns would indicate the existence of several modes at a finite plate angle. The patterns shown in Fig. 2 were made at constant exposure and are therefore greatly overexposed by the increased intensity with parallel plates. The half-intensity width of the parallel plate output is less than one minute of arc.

The phase variation over the area of the beam was studied by placing a double or multiple slit in the beam and recording the diffraction pattern. A nine-slit mask of 0.050-in. spacing and 0.002-in. slit width covering the full 0.45-in. aperture of the beam results in a diffraction pattern shown in Fig. 3. The minima are very low, indicating small phase variation over the aperture. A double-slit pattern at 0.0175-in. is also shown.

Since the linewidth of the maser oscillation is many orders of magnitude narrower than the resolution of the best available spectrometers or interferometers, standard optical techniques could not be used to measure the linewidth. Of the several possible methods examined, the most practical was found to be one based on analysis of the Fourier spectrum observed through a photomultiplier tube of the maser output. The photomultiplier is essentially a perfect square-law detector and should respond to beat frequencies. Hence, one can observe the spectral distribution obtained by beating all of the components of the line. The exact width of the original line determined from this distribution is independent of detailed assumptions on the line shape to within a factor of about two, providing the original line is not multiply peaked. In practice,

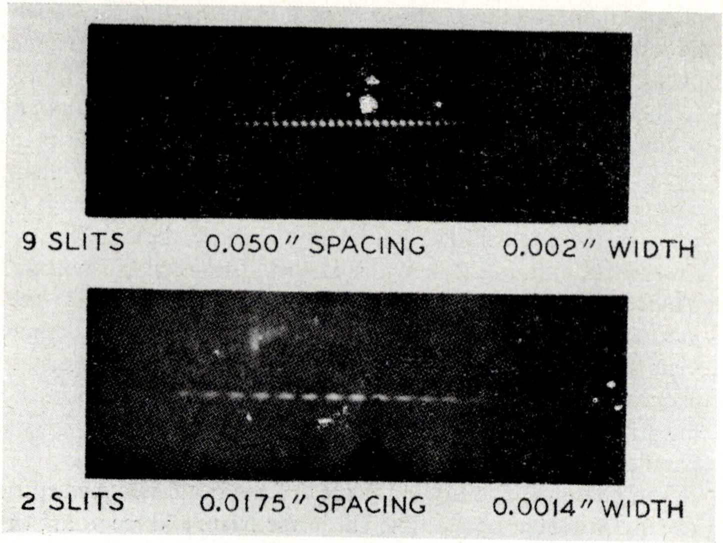

FIG. 3. Far-field diffraction patterns obtained by inserting multiple slits in the maser beam.

this spectral distribution will be superimposed on top of the flat spectrum obtained from shot noise in the photomultiplier tube, and the latter may be used to check the bandwidth of the system,

The system was first checked by observing the spectrum from an intense line emitted in an ordinary discharge tube using a spectrum analyzer. The latter was flat within the frequency range of our analyzer, covering from 100 cps to 13 Mc/sec. Using a grating spectrometer to isolate the strong line at 11 530 Å, the maser output was next examined. Our initial observations indicated a strong signal starting at zero frequency and extending to a half-width of about 100 kc/sec. Further examination showed the presence of an additional series of anharmonically related beats centered about frequencies other than zero. The location of these beats depended critically on the Fabry–Perot plate alignment. As the angular alignment of the plates was changed from 0 to 3 seconds, the beats became more widely spaced and weaker in intensity. Several possible spurious sources of beats were carefully examined and found not to be responsible for the observed signal. The possibility of "blocking oscillation" of the maser giving rise to an amplitude modulation of the output signal can be ruled out by the following observation: the peak frequency of a beat note is only a function of the plate adjustment and is independent of the factor by which the maser gain exceeds the oscillation threshold. The most plausible explanation found was that the beats arose from simultaneous oscillation in several different modes separated by frequencies in the Mc/sec range. Such a conclusion is not in disagreement with detailed calculations of the oscillatory modes by Fox and Li.[10] A distinct correlation was found between the visual structure of the beam observed through the infrared image converter, and the presence of the beats as the angular alignment of the plates was varied. Figure 4(a) shows several beat notes ranging from 50 kc/sec to 2 Mc/sec. Figures 4(b) and 4(c) show the structure of these beats on an expanded scale. It should be noted that in our present system the angular adjustment of the plates is also accompanied by a slight change in plate separation. The important conclusion to be drawn here regards the linewidth of the maser oscillation. As is evident from Figs. 4(b) and 4(c), the beat notes imply linewidths in the range from 10 to 80 kc/sec.

The use of excitation transfer for production of inverted population was suggested[1,2] by one of the authors. The He–Ne mixture described above is the first gaseous system which has led to maser oscillations at optical frequencies. Recently some evidence for the presence of inverted populations in a Hg–Zn mixture has been reported[11] which supports the wide applicability of this principle.[1,2]

In the course of this research, a number of co-workers have contributed in the design, construction, and measurement. We are indebted to F. Muller, P. S. Kubik, A. R. Strnad, D. MacNair, E. Koch, G. J. Wolff and G. E. Reitter. Dr. A. F. Turner of Bausch & Lomb has been very helpful in supplying the special

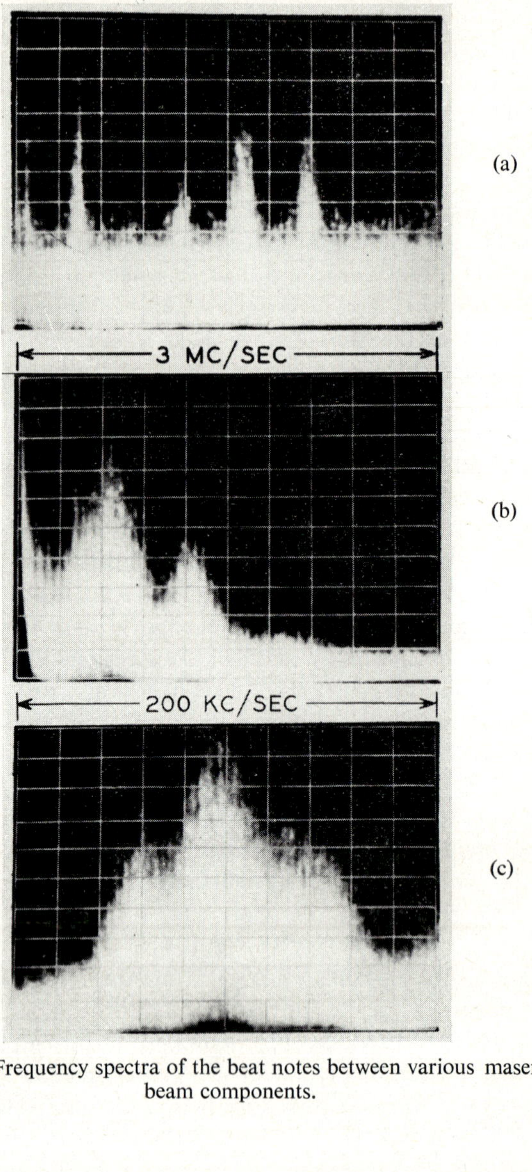

FIG. 4. Frequency spectra of the beat notes between various maser beam components.

evaporated reflectance films. In particular we would like to acknowledge the expert and patient assistance given by Mr. E. A. Ballik in the later stages of this experiment.

References

1. A. Javan, *Phys. Rev. Letters* **3,** 518 (1959).
2. A. Javan, in *Quantum Electronics*, edited by C. H. Townes (Columbia University Press, New York, 1960).
3. For simplicity, Paschen notation is used to describe the Ne energy levels.
4. Not all of the $1s$ levels are truly metastables: the $1s_2$ and $1s_4$ undergo transitions to the ground state; however, due to the trapping of the resonant ultraviolet photons these levels are effectively long-lived and quasi-metastable.
5. M. Pahl and U. Weimer, *Z. Naturforsch.* **12a,** 926 (1957).
6. A. V. Phelps, *Phys. Rev.* **99,** 4, 1307 (1955).
7. W. R. Bennett, Jr., A. Javan and E. A. Ballik, *Bull. Am. Phys. Soc.* **5,** 496 (1960).
8. R. Ladenberg, *Revs. Modern Phys.* **5,** 243 (1933).
9. A. L. Schawlow and C. H. Townes, *Phys. Rev.* **112,** 1940 (1958).
10. A. G. Fox and P. Li, *Proc. Inst. Radio Engrs.* **48,** 1904 (1960).
11. V. K. Ablekov, M. S. Pesiu and I. L. Fabelinski, *Zhur. Eksp. i Teoret. Fiz.* **39,** 892 (1960).

10

Laser Conditions in Semiconductors*

Maurice G. A. Bernard and Georges Duraffourg

Centre National d'Etudes des Télécommunications, Issy-les-Moulineaux, Seine

Summary

The possibility of obtaining stimulated emissions in semiconductors has been considered for transitions between the conduction band and the valence band, or between one band and an impurity level. If the occupation of the bands and of the impurity levels is taken into account by quasi-Fermi levels the necessary condition for stimulated emission to be possible turns out to be simply: $\Delta F > h\nu$ where ΔF is the difference of the quasi-Fermi levels of the initial and final state, and ν is the emitted frequency. The existence of such quasi-Fermi levels is discussed, and it is shown that the above condition is due to the second law of thermodynamics. Direct interband transitions in InAs or InSb, and transitions between the conduction band Zn and In acceptor levels, respectively, in Ge and Si are thought to be sufficiently attractive to be studied experimentally.

Many papers[1] have been recently published on lasers but, most of them consider only stimulated emission between localized levels. Thus it is well known that the necessary condition to obtain amplification of electromagnetic waves by stimulation of emitted radiation is the so called "population inversion" between the two localized levels of the transition. However in 1958, P. Aigrain[2] has shown that this condition was not necessary in the case of the emission of two bosons, one being stimulated the other thermalized.

* *Physica Status Sol.* **1,** 699 (1961).

[1] For a recent bibliography on lasers see, for instance, review articles by A. Authier (to be published in *NUCLEUS*) and by N. G. Basov, O. N. Krokhin and Yu. M. Popov (1).

Actually, in the one-particle approximation an electronic state in a solid may be represented by a Bloch wave defined over the whole crystal: let us assume two such states: one with a wave vector k_i and energy $E_v(k_i)$, the other with a wave vector k_j and energy $E_c(k_j)$. Let us suppose that the first of these states belong to a group of states (v) and the second to a group of states (c); we shall first suppose that these two groups are respectively the valence band and the conduction band of a semiconductor.

Under equilibrium conditions the occupation probability of any state of energy $E(k)$ is given by the Fermi–Dirac function

$$f = \frac{1}{1 + \exp\left[E(k) - F_0\right]/k_0 T}$$

where k_0 is the Boltzmann constant and F_0 the Fermi level or chemical potential for electrons. If the crystal is no longer in equilibrium, W. Shockley [2] has shown that, under certain conditions, which we shall discuss later, the occupation probability of any state of the conduction band is given by

$$f_c = \frac{1}{1 + \exp\left[E(k) - F_c\right]/k_0 T}$$

where F_c is the "quasi Fermi level" for the electrons of the conduction band; in the same way we can define a "quasi Fermi level" F_v for the holes of the valence band; at equilibrium $F_c = F_v = F_0$.

A state $E_v(k_i)$ of the valence band is in general connected to a state $E_c(k_j)$ of the conduction band by a direct radiative transition; let W_v^c be the probability per unit time of such a process; in a radiation field containing a density $P(\nu)$ of photons of energy $h\nu$,

[2] P. Aigrain, unpublished lecture at the "International Conference on Solid State Physics in Electronics and Telecommunications", Bruxelles 1958. The same idea has been suggested recently and independently by N. G. Basov, O. N. Krokhin and Yu. M. Popov, Second International Conference on Quantum Electronics.

the number N_a of quanta absorbed per unit time is

$$N_a = AW_v^c f_v(k_i)[1 - f_c(k_j)]P(v) \tag{1}$$

The number N_e of quanta emitted per unit time by stimulated emission is

$$N_e = AW_c^v[1 - f_v(k_i)]f_c(k_j)P(v) \tag{2}$$

where the proportionality coefficient A includes the densities of states of the valence band and of the conduction band; taking into account $W_v^c = W_c^v$, the necessary condition for amplification to be possible, $N_e > N_a$, can be written

$$f_c(k_j)[1 - f_v(k_i)] > f_v(k_i)[1 - f_c(k_j)]$$

that is to say

$$\exp (F_c - F_v)/k_0T > \exp [E_c(k_j) - E_v(k_i)]/k_0T \tag{3}$$

and as

$$E_c(k_j) - E_v(k_i) = hv$$

the above conditions simply reduces to

$$F_c - F_v > hv. \tag{4}$$

This condition can be generalized in different ways. First, let us consider an indirect transition in which one or several phonons are emitted or absorbed; the expressions (1) and (2) should be modified; by a straightforward calculation it is shown that, if the phonons are thermalized, the inequality (3) is replaced by

$$\exp (F_c - F_v)/k_0T > \exp [E_c(k_j) - E_v(k_i) - \sum_q \varepsilon_q \hbar\omega_q]/k_0T \tag{5}$$

where $\hbar\omega_q$ is the energy of the qth phonon involved and where $\varepsilon_q = \pm 1$ ($+1$ if the phonon is emitted during the stimulated process and -1 if the phonon is absorbed); but the frequency of the radiation we want to stimulate is given by

$$hv = E_c(k_j) - E_v(k_i) - \sum \varepsilon_q \hbar\omega_q$$

and again the laser condition is given by relation (4).

Secondly, let us consider a radiative transition, direct or indirect, between one of the bands, say, for instance, the conduction band, and a given impurity level of the forbidden energy gap. The statistics of occupation of defect centers of different charge conditions and various degeneracies [3] and the statistics of recombination on such centers [4, 5] have been treated by different authors; an equivalent treatment has been made by one of us [5] using the concept of a "quasi Fermi level" associated with each impurity level; let N_s be the number of centers in charge condition s, N_{s+1} the number of centers in charge condition $s + 1$; the energy $E_{s+1/2}$ of the level $(s, s + 1)$ is defined by [4, 5]

$$E_{s+1/2} = k_0 T \log Z_s / Z_{s+1}$$

where Z_s and Z_{s+1} are two partition functions.

At thermal equilibrium we have

$$N_{s+1}/N_s = \exp{(F_0 - E_{s+1/2})/k_0 T} \tag{6}$$

where F_0 is the Fermi level. Under non-equilibrium conditions the ratio N_{s+1}/N_s is no longer equal to that given by equation (6); under certain conditions which we shall discuss later, we may define a "quasi-Fermi level" $F_{s+1/2}$ for the level $(s, s + 1)$ by the relation

$$N_{s+1}/N_s = \exp{(F_{s+1/2} - E_{s+1/2})/k_0 T}$$

Now a simple calculation shows that the laser condition for the transition between a state of the conduction band $E_c(k_j)$ and a state of the defect level $(s, s + 1)$ is

$$F_c - F_{s+1/2} > h\nu \tag{7}$$

which is again equivalent to relation (4).

Thus, under the assumption that, for different groups of electronic states in a semiconductor (conduction band, valence band, impurity levels) we may define a "quasi Fermi-level" associated with each group, the necessary condition for electromagnetic waves to be amplified by stimulation of emitted radiation is given by the simple formula (4). The concept of a "quasi Fermi level" for a group of states is valid if the electrons of the group exchange energy with a heat reservoir, or thermostat, in a time short compared to the time necessary for the population to come back to equilibrium by transitions between states of the different groups; this is apparently the case in most semiconductors. In other words, we may look at any semiconductor laser as a collection of a very large number of three-level masers (or even more than three) having nearly the same signal frequency but a wide spectrum of "idler frequencies"; to admit the existence of a "quasi Fermi level" is equivalent to express that the corresponding "idler frequencies" are strongly coupled to a thermostat. If the pump transitions are provided, through a filter, by a good contact of the semiconductor with the radiation of a black body at a temperature T_p, a simple calculation [6], based upon a detailed balance argument applied to the continuous pumping action, shows that

$$F_c - F_v = h\nu_p\left(1 - \frac{T}{T_p}\right) = h\nu_p\eta_c$$

where $\eta_c = 1 - (T/T_p)$ is the Carnot efficiency of a heat engine working between temperatures T and T_p; but the efficiency of the laser is $\eta = h\nu/h\nu_p = \nu/\nu_p$ and the laser condition (4) can now be rewritten

$$\eta < \eta_c$$

which shows that the efficiency of a heat engine giving coherent light must be less than the Carnot efficiency.

The interest of such an approach is emphasized by the following remarks:

(i) The actual operation of a semiconductor laser needs not only the necessary condition (4) to be fulfilled but also requires the net emission to be greater than the losses: diffraction, reflexion, free carrier absorption etc . . . But this sufficient condition should be calculated in each particular case; thus relation (4) is a useful step in predicting what are the most favourable cases to be more carefully studied.

(ii) For a direct transition, $h\nu = E_G$ (where E_G is the width of the energy gap) and condition (4) requires that the number of carriers injected be so large that both the electron and hole populations are degenerate or nearly degenerate; low effective mass materials at low temperature should be favourable; as an example it is proposed to use, say, an n type InSb or InAs crystal at liquid helium temperature with a pump transition ν_p between the top of the valence band and the Fermi surface in the conduction band and a stimulated emission ν_s between the bottom of the conduction band and the top of the valence band[3] (Fig. 1).

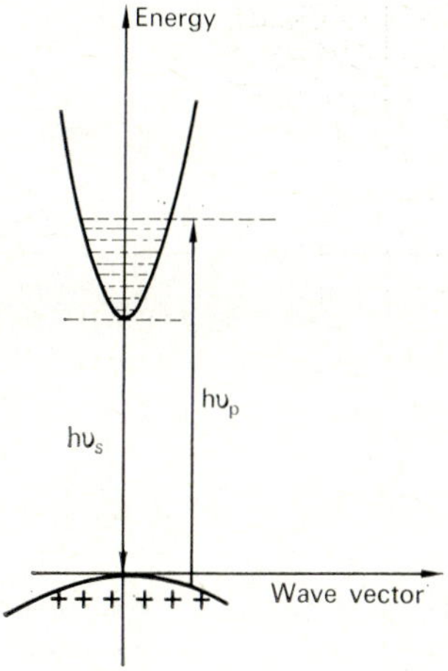

FIG. 1. Pump and signal transitions in the band structure of InSb or InAs (degeneracies in the band structure have been omitted for simplicity).

(iii) For an indirect transition $h\nu$ is less than the energy gap

$$h\nu = E_G - \hbar\omega$$

where $\hbar\omega$ is the energy of an emitted phonon; this must reduce greatly the pump power necessary for laser effect; we find a result which had been anticipated by Aigrain who was the first, to propose in 1958 the indirect recombination radiation in Germanium to obtain an emission of coherent light.[2]

[3] The same proposal has also been made independently by C. Benoît à la Guillaume and C. Tric.

N

(iv) For an impurity transition, again, $h\nu$ is less than E_G; let us consider, for instance, the radiative transition on the second acceptor level of Zn in Germanium (Fig. 2): the capture of an electron of the conduction band by a Zn^- ion is anticipated to be a slow radiative process while the capture of a hole by a Zn^{--} ion should be very fast; thus the quasi Fermi level for electrons F_c

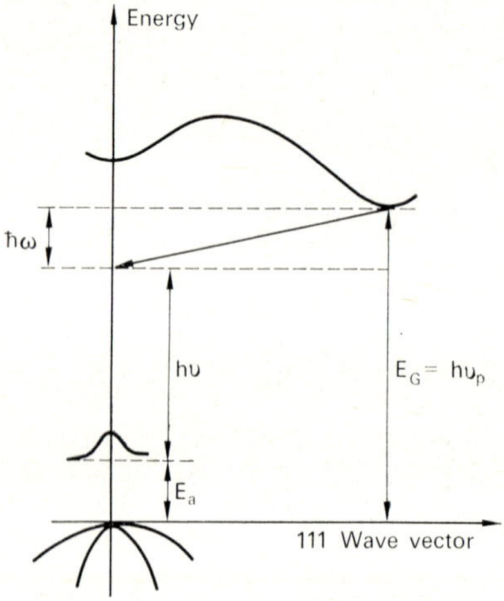

FIG. 2. Band structure of Ge with the different involved transitions. The Fourier transform of the wave function of the 2nd acceptor level of zinc is estimated to be important only in the vicinity of $k(000)$; the first acceptor zinc level, which should also offer interesting possibilities has been omitted for clarity.

will rise up with injection while the "quasi Fermi level" $F_{3/2}$ for the Zn level will be close to the "quasi Fermi level" for holes F_v and thus close to the valence band. On the other hand the energy of the emitted photon $h\nu$ is much less than the energy gap and given by

$$h\nu = E_G - E_a - \hbar\omega$$

where E_a is the ionisation energy of the 2nd acceptor level of Zn ($E_a = 0{\cdot}09$ eV) and $\hbar\omega$ is the energy of a phonon emitted by the radiative transition which should be indirect.

Another interesting case is that of Indium in Silicon; according to recent data [7] the capture of an electron on a neutral Indium atom is a slow radiative process while the capture of a hole on a negatively charged Indium atom is faster by six orders of magnitude.

A detailed paper on the subject will soon be published; experiments on InSb, InAs, Zn in Ge and In in Si are under progress.

References

1. N. G. Basov, O. N. Krokhin and Yu. M. Popov, *Soviet Physics Uspekhi* **3,** 702 (1961).
2. W. Shockley, *B.S.T.J.* **28,** 435 (1949).
3. P. T. Landsberg, *Proc. Phys. Soc.* **B 69,** 661 (1956).
   C. H. Champness, *Proc. Phys. Soc.* **B 69,** 1335 (1957).
   W. Shockley and J. T. Last, *Phys. Rev.* **107,** 392 (1957).
4. R. N. Hall, *Phys. Rev.* **83,** 228 (1951) and **87,** 387 (1952).
   W. Shockley and W. J. Read, *Phys. Rev.* **87,** 835 (1952).
   P. T. Landsberg, *Proc. Phys. Soc.* **B 70,** 282 (1957).
   M. Bernard, *J. of Electronics and Contr.* **4,** 15 (1958).
   C. T. Sah and W. Shockley, *Phys. Rev.* **109,** 1103 (1958).
5. M. Bernard, Défauts dans les semiconducteurs, Cours du 3e cycle de Physique des Solides, Faculté des Sciences d'Orsay.
6. H. E. D. Scovil and E. O. Schulz-Dubois, *Phys. Rev. Letters* **2,** 262 (1959).
7. Y. E. Pokrovsky and K. J. Svistunova, International Conference on Photoconductivity, Ithaca, USA, Aug. 1961.

11

Coherent Light Emission from GaAs Junctions*

R. N. Hall, G. E. Fenner, J. D. Kingsley, T. J. Soltys, and R. O. Carlson

General Electric Research Laboratory, Schenectady, New York

Coherent infrared radiation has been observed from forward biased GaAs *p–n* junctions. Evidence for this behavior is based upon the sharply beamed radiation pattern of the emitted light, upon the observation of a threshold current beyond which the intensity of the beam increases abruptly, and upon the pronounced narrowing of the spectral distribution of this beam beyond threshold. The stimulated emission is believed to occur as the result of transitions between states of equal wave number in the conduction and valence bands.

Several requirements must be fulfilled[1] in order that such stimulated emission can be observed: (a) the electron and hole populations within the active region must be large enough that their quasi-Fermi levels are separated by an energy greater than that of the radiation; (b) losses due to absorption by other processes must be small relative to the gain produced by stimulated emission; and (c) the active region must be contained within a cavity having a resonance which falls in the spectral range within which stimulated emission is possible.

In our structure, the necessary population inversion is produced by injection of carriers from the degenerate *n*- and *p*-type end

* *Phys. Rev. Letters*, **9**, 366 (1962).

regions of the junction into the transition region. Condition (b) is the most difficult to fulfill, since the stimulated radiation propagates in the plane of the junction where the active region may be only a fraction of a wavelength in thickness and has a wave front which laterally extends many wavelengths into the passive *n*- and *p*-type regions bounding the junction plane. Laser action is favored by the large matrix element for band-to-band radiative recombination compared with that for free carrier absorption in GaAs and by the fact that the energy of the emitted radiation is below the absorption threshold of the degenerate material bounding the junction.

The diodes can be described approximately as cubes 0.4 mm on an edge, the junction lying in a horizontal plane through the center. Current is passed through the junction by means of Ohmic contacts attached to the top and bottom faces. The front and back faces are polished parallel to each other and perpendicular to the plane of the junction, in order to satisfy condition (c) above. Current is applied in the form of pulses of 5 to 20 μsec duration with the diode immersed in liquid nitrogen.

We have studied the radiation pattern by means of an infrared convertor tube which provides an image that can be examined visually or photographically. Examples of such patterns are given in Fig. 1. The first two photographs show the pattern produced by a diode with the current below and above threshold, with the image convertor 6 cm from the diode. Figure 1(c) shows the same diode above threshold with the screen 15 cm away, so that the radiation pattern is correspondingly enlarged. In these photographs the horizontal bands result from interference between the light produced at the junction and its image in the metal disc on which the diode was mounted, and their presence does not imply coherence of the source. On the other hand, the vertical bands which appear only in Figs. 1(b) and 1(c) are due to interference between the waves emitted at various points along the edge of the junction at the front face of the cube, and their appearance is evidence that a definite phase relation exists between the light emitted at various points along the junction, i.e., that the light is coherent. The angular

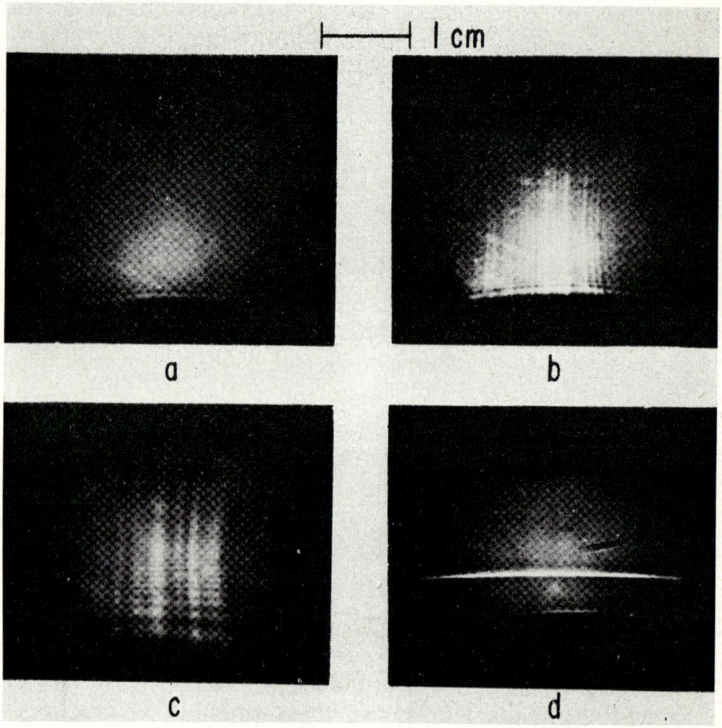

FIG. 1. Radiation patterns observed with image tube a distance d from junction. (a) and (b). Diode L-69 below and above threshold, $d = 6$ cm. (c) Diode L-69 above threshold, $d = 15$ cm. (d) Diode L-75, $d = 5$ cm.

separation between the interference maxima is consistent with the dimensions of the junction and the wavelength of the radiation. However, the relative intensities of the maxima produced by this and other similarly constructed diodes indicate that the radiation is not produced by the most elementary mode of this type of cavity.

Measurements were made of the light intensity from diode L-70 as a function of junction current at 8420 Å on the axis of the beam. Below 5000 A/cm^2 the light intensity varied linearly with current

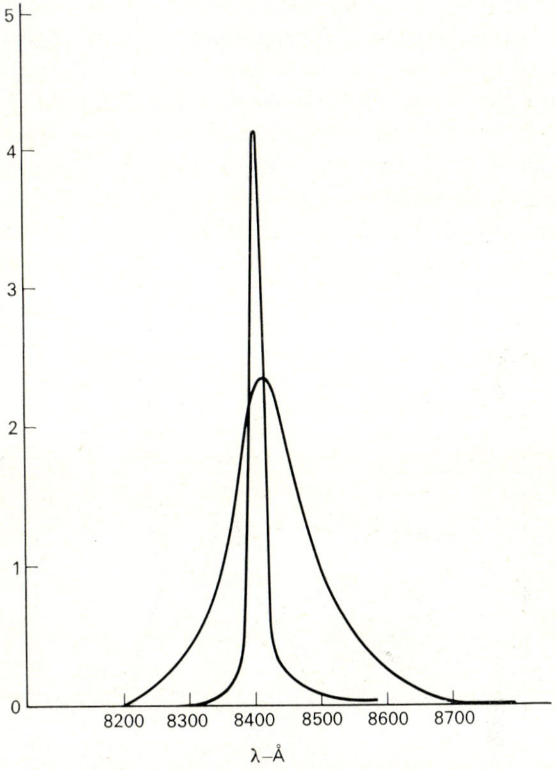

FIG. 2. Spectral distributions from diode L-69 below and above threshold. Different vertical scales.

density. Near 8500 A/cm^2 the intensity increased very rapidly with current, reaching a value about ten times the extrapolated low-current intensity at 20,000 A/cm^2. Such a current threshold is characteristic of the onset of stimulated light emission, and it is significant that the azimuthal interference maxima of Fig. 1 make their appearance at this threshold.

Figure 2 shows the spectral distribution measured with the spectrometer located on the beam axis at several values of current density. Below threshold, the spectral width at half maximum is

125 Å, in agreement with the measurements of Keyes and Quist.[2] As the current is increased through threshold, the spectral width decreases suddenly to 15 Å, in a manner which again is characteristic of the onset of stimulated emission. If this 15 Å width is due to a single resonant mode so that the emission is homogeneously broadened by the lifetime of this mode, then the oscillations persist only for a period of 1.6×10^{-12} sec or for a time long enough for the radiation to travel 0.1 mm, which is less than the length of

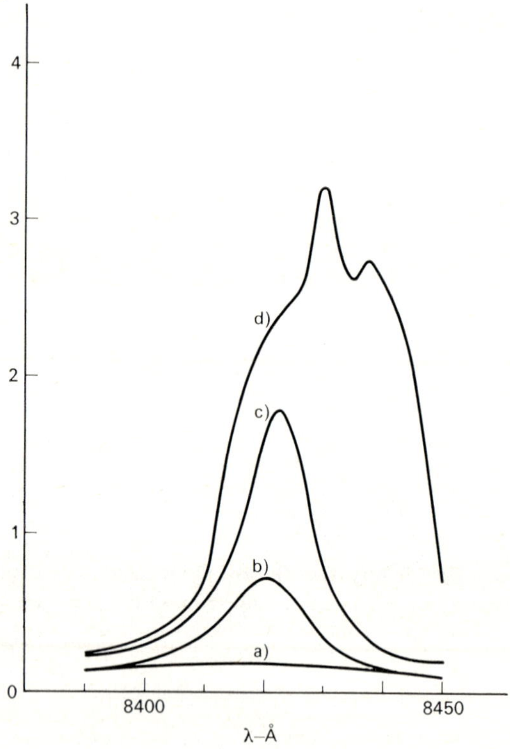

FIG. 3. Spectral distributions of diode L-70 at various currents, same vertical scale for all curves, arbitrary units. (a) 6000 A/cm², (b) 8600 A/cm², (c) 10,400 A/cm², (d) 20,000 A/cm².

the crystal. However, the broadening may instead be due to oscillations occurring at several unresolved modes during the pulse or to a variation in the wavelength of the resonant mode during a pulse. The instrumental resolution was about 3 Å.

Additional modes appear at longer wavelengths when the current is further increased beyond threshold, as illustrated in Fig. 3. The fact that the additional modes appear at lower energy suggests that the principal dissipative process becomes more effective at longer wavelengths, as would be the case for free carrier absorption. The separation of these additional modes is about three times that calculated on the assumption that each corresponds to one additional half wavelength between the plane surfaces, which were 0.32 mm apart in this diode.

Other junctions have been constructed which exhibit quite different characteristics. These produce a radiation pattern which is almost uniform in the azimuthal plane, but which is only a few tenths of a degree wide in the vertical plane, as shown in Fig. 1(d). This pattern implies that there is coherence over a distance of the order of 100 μ in the vertical direction but virtually no spatial coherence in the horizontal direction.

While stimulated emission has been observed in many systems, this is the first time that direct conversion of electrical energy to coherent infrared radiation has been achieved in a solid state device. It is also the first example of a laser involving transitions between energy bands rather than localized atomic levels.

References

1. M. G. A. Bernard and G. Duraffourg, *Physica Status Solidi* **1,** 699 (1961).
2. R. J. Keyes and T. M. Quist, *Proc. Inst. Radio Engrs.* **50,** 1822 (1962).

12

Hole Burning Effects in a He–Ne Optical Laser*

W. R. BENNETT, JR.

Bell Telephone Laboratories, Murray Hill, New Jersey

Summary

A study has been made of pulling effects by the amplifying media on the TEM_{00} modes of a helium–neon maser using a circular plane mirror Fabry–Perot cavity in which the mirror separation was known with precision. Approximate expressions are derived for mode pulling in homogeneously and inhomogeneously broadened optical masers. The experimental results suggest that the losses in the maser are not determined entirely by the mirror reflectance coefficient. A power-dependent splitting of the beat frequencies between simultaneously oscillating modes is explained in terms of a non-linear frequency-dependent pulling effect arising from inhomogeneous broadening. The case of Lorentzian holes burned in a Gaussian line is treated specifically. It is suggested that the anomalous variation of beat frequencies with pumping rate results from hole repulsion effects and that the hole widths required arise from the combined effects of small-angle elastic scattering and stimulated emission. It is a consequence of the interpretation that the number of simultaneously oscillating even-symmetric modes in the helium–neon maser may easily be determined from the Fourier spectrum of the beat frequencies. A method is described by which the central cavity resonance may be stabilized near the center of the Doppler line in the case of three oscillating symmetric modes.

1. Introduction

The narrow linewidths inherent in the helium–neon optical maser[1,2] are suggestive of a number of basic experiments.

* *Phys. Rev.* **126**, 580 (1962).

[1] A. Javan, W. R. Bennett, Jr. and D. R. Herriott, *Phys. Rev. Letters* **6**, 106 (1961).

[2] A. Javan, E. A. Ballik, and W. L. Bond (to be published) have recently remeasured the linewidths in the first maser and have found that they are inherently in the order of 2 cycles/sec or less.

However, the ultimate usefulness of this narrow source of radiation, both as a research tool and a communications device, is largely limited by the frequency stability that may be obtained. For the latter reason, the present investigation was undertaken. Since the oscillation frequencies are primarily determined by the cavity resonances, and therefore the dimensions of the maser, it was not obvious *a priori* that a long-term frequency stability of much better than half the separation between adjacent cavity resonances (about 80 Mc/sec) could actually be obtained. It became apparent during the course of this research, however, that a method existed by which one could both determine the number of simultaneously oscillating modes and, in the case of three oscillations, set the central mode near the center of the Doppler line. The absolute stability that may be achieved in this way is still unknown and is currently under investigation. It seems clear that the method may at least be used to obtain an extremely high degree of relative stability.

2. Maser Construction

For the purpose of making the present study, the He–Ne maser shown in Fig. 1 was constructed. The mechanical design of the interferometer was altered from that of the original He–Ne maser,[1] in order to minimize mechanical fluctuations and permit an accurate determination of the plate separation. Each of the Fabry–Perot plates was fastened by spring clips to internal three-point

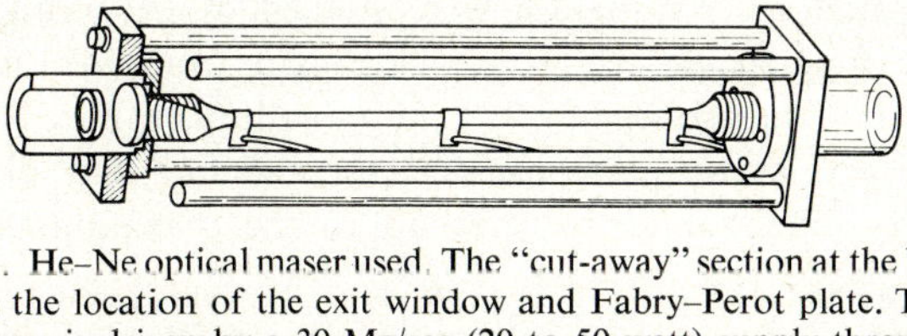

FIG. 1. He–Ne optical maser used. The "cut-away" section at the left shows the location of the exit window and Fabry–Perot plate. The discharge is driven by a 30-Mc/sec (20 to 50 watt) supply through the three external electrodes shown. The plate separation and angular alignment is controlled magnetostrictively by coils (not shown) placed about each of the four Nilvar supporting rods.

mountings which, in turn, were securely fastened to the large external flanges shown at either end of the maser. The large flanges were surface-ground and separated by four 1-in. diameter Nilvar rods, cut to lengths which were identical within 0.0005 in. The bellows incorporated at each end of the discharge tube were sufficiently flexible to permit changing the plate separation by about 4.5% through the insertion of accurately machined spacers at the ends of the 1-in. diameter rods. Of several methods tried for controlling the plate angular alignment, one based on the magnetostrictive effect in the Nilvar supporting rods was found to be the most satisfactory. The magnetostrictive effect was also found to present a convenient method for changing the plate separation; however, with the adjustments used, some coupling between the plate separation and angular alignment controls existed. A more detailed description of the alignment method and interferometer construction will be given elsewhere. The improved mechanical stability obtained with the present maser arises largely by virtue of the fact that the tuning is accomplished through the slight distortion of a highly rigid structure. In other respects, the maser had properties similar to those reported in the first He–Ne maser with the exception that the device oscillated only on the strongest ($2s_2$ to $2p_4$ in Paschen notation at 11,522.76 A)[3] of the five transitions of neon previously observed. The plates used were flat to $\cong \lambda/100$ over the beam diameter and had a reflectance of 99% at the wavelength used prior to insertion in the maser.[4]

3. Experimental Results

The modes of oscillation were studied by observing the difference frequencies produced when light from the maser was focussed on the surface of a 7102 photomultiplier tube. The photosurface acts as a square-law detector and one may observe all possible difference frequencies up to the maximum frequency

[3] The author is indebted to D. L. Wood for his help in firmly establishing the identity of this line. The small possibility that the strongest maser transition might have been the $2s_4$–$2p_7$ line at 11,525.016 Å had not previously been ruled out experimentally.

[4] The author is indebted to D. R. Herriott for making these measurements.

response characteristic of the tube. This effect was first observed by Forrester, Gudmundsen, and Johnson[5] using incoherent light sources and more recently was used to investigate the linewidths obtained in the first helium–neon maser.[1,2] A previous report on the difference-frequency spectrum obtained in this way from a He–Ne maser has been given by Herriott[6] and, for the sake of clarity, will be summarized here. Since the single-pass gain obtainable from a He–Ne maser is small, one is forced to use a Fabry-Perot cavity with an enormously high Q in order to exceed the threshold requirement for oscillation. In practice this implies a cavity width which is much narrower than the Doppler width for one transition and that the frequencies of oscillation will be primarily determined by the cavity resonances. The dominant cavity resonance frequencies are determined by requiring that the cavity length L be a half-integral multiple of the wavelength. These resonant frequencies correspond to the first even-symmetric radial modes calculated by Fox and Li[7] and differ in frequency by $c/2L \cong 160$ Mc/sec for the present work, where c is the velocity of light in vacuum. Since these frequency separations are small compared to the Doppler width, the maser may oscillate on several of these frequencies simultaneously. In addition to the even-symmetric modes, Fox and Li have shown that the next longitudinal modes of importance are those possessing odd radial symmetry and that these differ from the former by frequencies of the order of a Mc/sec with the present geometry. Hence, as was reported by Herriott,[6] one obtains a difference frequency spectrum of the type shown in Fig. 2. That is, a peak at zero frequency corresponding to each line beating with itself, followed by a peak $\cong 1$ Mc/sec away corresponding to all possible differences between the first even- and odd-symmetric radial modes having the same value of $c/2L$; a peak at $c/2L$ corresponding to the differences between all

[5] A. T. Forrester, R. A. Gudmundsen and P. O. Johnson, *Phys. Rev.* **99**, 1691 (1955).

[6] D. R. Herriott, in *Advances in Quantum Electronics*, edited by J. R. Singer (Columbia University Press, New York, 1961), p. 49.

[7] A. G. Fox and Tingye Li, *Bell System Tech. J.* **40**, 453 (1961).

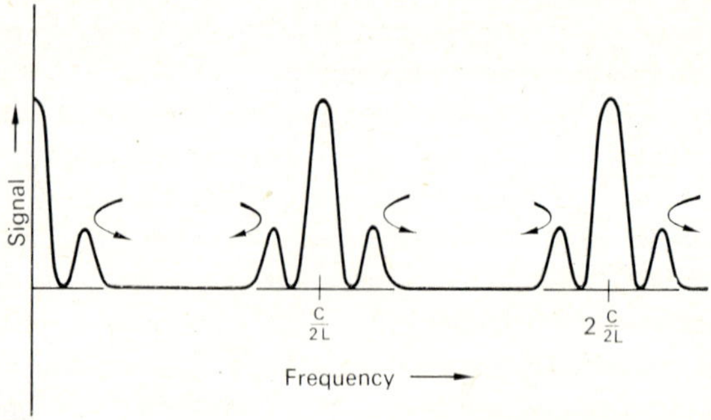

FIG. 2. [illegible] fr[illegible]ossible difference frequencies in [illegible]r.

even-symmetri[illegible] $c/2L$, and all odd-symmetric modes separate[illegible]ed by two satellites corresponding to the possible e[illegible] and odd-even difference frequencies at approximately $c/2L$, etc. As indicated by the arrows in Fig. 2, the satellites obtained as a result of the beats between the odd-even symmetric modes vary strongly with the interferometer plate alignment, reaching a minimum separation from the $c/2L$ beats at parallel alignment. For the latter reason, they are both easily identifiable, and unsuitable for the present study. The discussion which follows is concerned entirely with the central peaks obtained at mulitiples of $c/2L$ corresponding to the differences between the even-symmetric modes. That is, the low power data were taken under conditions where the satellites about the $c/2L$ and $2(c/2L)$ beats (and therefore the odd-symmetric modes) were absent.

The light output of the maser was found to have more or less random polarization. The output varied from the extreme case of linear polarization with arbitrary orientation of the electric field vector to nearly perfect circular polarization over time intervals in the order of seconds. This result would be expected from ideal cylindrical geometry and is in contrast with the behavior of the first He–Ne maser.[1,8] In addition, the various modes of oscillation in the present maser were also found to have unrelated polarizations. The latter was established by examining the behavior of the beat frequencies when a linear polaroid was inserted in the beam. Javan and Ballik[9] have observed that the central $c/2L$ beat may sometimes not be observable without the use of a polaroid. Their observation may be interpreted through the assumption that the two modes producing the beat are polarized linearly at right angles. The phototube, being a reasonably good square-law detector, fails to respond to the beat since the difference frequency is contained in the scalar product of the two electric fields. The insertion of a polaroid at 45° to the field components, however, restores the beat. That [illegible] transmits components of the two fields which are parallel. In order to avoid missing any of the beats in the present work, a polaroid was always inserted in the beam and [illegible] was obtained in which all beats present were [illegible] comparable amplitude. For this reason, the relative amplitudes (for example of the split components in Fig. 3) are arbitrary.

The first measurements were made on the absolute frequency of the central peak corresponding to a difference frequency of $c/2L$. These measurements were made under the conditions of low power in which this beat was not split and the beat at $2c/2L$ was not observable, i.e., the conditions under which the data in Fig. 3(a) were taken. Measurements were made at three discrete interferometer lengths and these data are shown in Table 1. The uncertainty in the absolute frequency measurement ($\cong 20$ kc/sec) arises primarily from the variation of the beat frequency with power and plate separation as discussed below. As is obvious from Table 1, the beat frequency is not equal to $c/2L$, but is less than $c/2L$ by about 1 part in 800. This effect may be readily explained on the basis of mode pulling by the Doppler line and is discussed in

[8] The first He–Ne maser was linearly polarized in a direction closely correlated with a striated pattern in the dielectric coatings. The absence of such striated patterns in the present maser may have resulted from the use of more modest bake-out temperatures.

[9] A. Javan and E. A. Ballik (private communication).

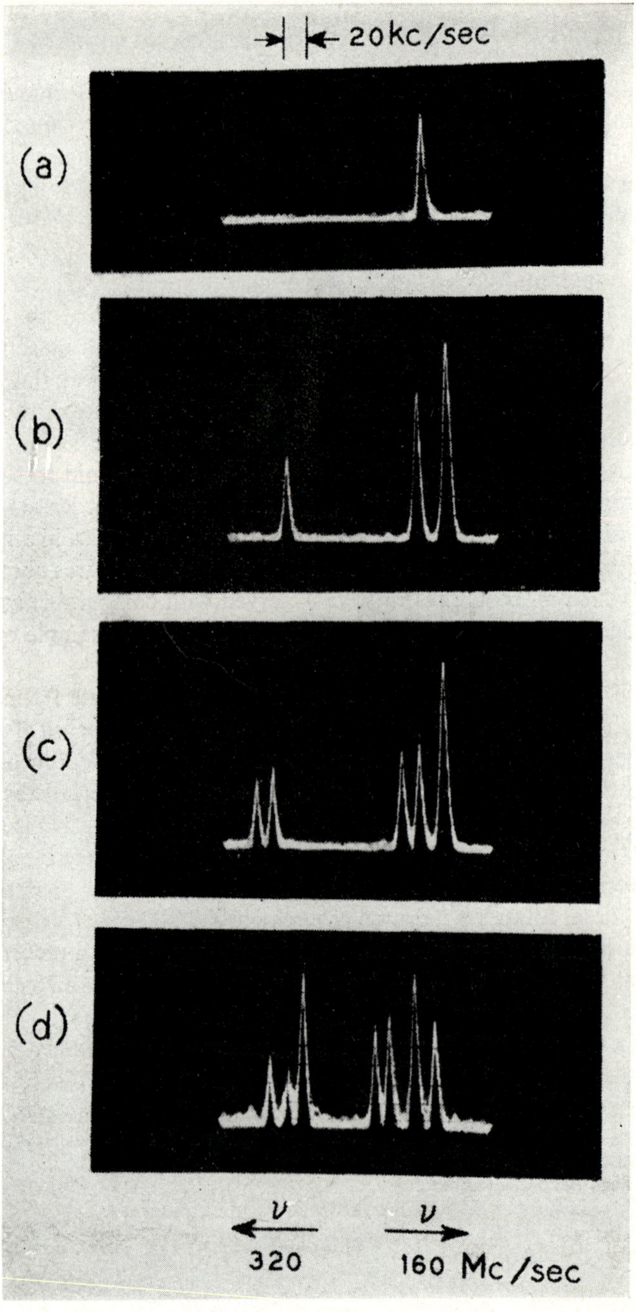

FIG. 3. Splitting of $c/2L$ and $2(c/2L)$ beats with power. Power increases from (a) to (d).

Sec. 13. It is a consequence of the analysis in Sec. 5 and the data in Table 1 that the loss in the interferometer exceeds the known reflectance loss by a factor of about two.

TABLE 1. Comparison of measured beat frequencies with the difference between calculated cavity resonance frequencies for three separate cavity lengths

$c/2L$ (in Mc/sec)	Measured beat (in Mc/sec)	Fractional difference between $c/2L$ and measured beat
161.316 ± 0.022	161.107 ± 0.010	$-(1.3 \pm 0.2) \times 10^{-3}$
158.713 ± 0.020	158.531 ± 0.010	$-(1.1 \pm 0.2) \times 10^{-3}$
156.189 ± 0.018	155.982 ± 0.010	$-(1.3 \pm 0.2) \times 10^{-3}$

There are two interesting aspects of the data obtained at increasing power levels:

(a) The frequency of the $c/2L$ beat was found to increase with increasing power. The magnitude of the increase before splitting occurred (see below) was dependent on the setting of the interferometer plates. It varied from nearly zero to a maximum of about 30 kc/sec. As will become evident from the analysis in Sec. 5, such an increase is anomalous. That is, for any singly-peaked line one would expect the pulling towards the line center to increase with the number of excited atoms and that the pulling would be less for a cavity resonance near the line center than for one farther away from it. Hence one would expect the frequency separation between adjacent cavity resonances to decrease with increasing power. The same anomalous behavior was encountered in the split components described below. The anomaly is explained through the hole repulsion effect in Sec. 13.

(b) A power-dependent splitting of the $c/2L$ and $2(c/2L)$ beats was encountered. Figs. 3(a) through 3(d) show data taken simultaneously on the central beats at both $c/2L$ and $2(c/2L)$ with increasing rf power. The data on the right show the $c/2L$ beat (160 Mc/sec) and those on the left the $2(c/2L)$ beat (320 Mc/sec). The frequency scale is the same in each case as indicated by the 20 kc/sec interval shown, with the exception that increasing frequency occurs in opposite directions for the two sets of data. These data were taken by adjusting the maser length so that the $c/2L$ beat occurred within the first *if* band of the spectrum analyzer used and by tuning the local oscillator in the analyzer to correspond to the $2(c/2L)$ beat. The data shown in Figs. 3(a) through 3(d) were taken at increasing power levels, where each figure represents about 10 oscilloscope traces. In Fig. 3(a) the $c/2L$ beat is present and the $2(c/2L)$ beat is not. As the power increases [Fig. 3(b)], the $c/2L$ beat splits by an amount which typically was in the order of 20 kc/sec and the $2(c/2L)$ beat appears. The separation between the two $(c/2L)$ components in Fig. 3(b) was dependent on the interferometer length. By changing the plate separation magnetostrictively, the splitting shown in Fig. 3(b) could be varied from a maximum of about 30 kc/sec to a minimum of $\leqslant 1$ kc/sec (i.e., 1 kc/sec was the minimum *if* resolution of the apparatus used). With a further increase in power [Fig. 3(c)], the $c/2L$ beat breaks into three components and the beat at $2(c/2L)$ splits. Again the frequency of the components was found to increase with power and the splitting was dependent on the interferometer setting. For some interferometer settings, the odd-symmetric modes (as evidenced by the appearance of the satellites in Fig. 2) were also present during conditions typical of Fig. 3(c). Their presence or absence seemed to have relatively little effect on the splitting of the central $c/2L$ and $2(c/2L)$ beats. Finally, at the highest power level shown in Fig 3, the $c/2L$ beat has broken into four components and the $2(c/2L)$ beat into three.

The data in Fig. 3 may be interpreted in the following way: Some nonlinear frequency-dependent pulling mechanism exists whereby the differences between adjacent oscillating modes are not identical. In Fig. 3(a) the gain is only sufficient for the two modes nearest the line center to break into oscillation. Consequently there is one component at $c/2L$ and none at $2(c/2L)$. In Fig. 3(b) the gain is adequate for three modes to oscillate; there exist two discrete-frequencies at $c/2L$ and one at $2(c/2L)$. In Fig. 3(c) four modes are oscillating; there exist three different ways of obtaining the $c/2L$ beat and two for the $2(c/2L)$ beat. Finally, in Fig. 3(d) the gain is high enough for five modes to oscillate; there exist four ways of obtaining the 160-Mc/sec beat and three ways of obtaining the 320 Mc/sec beat. The expected components at frequencies higher than 320-Mc/sec have since been observed.[10]

There are two important consequences of the interpretation of the data in Fig. 3. The first is that one may deduce the number of even-symmetric modes which are oscillating from the beat splitting. The second is that in the case of three simultaneously oscillating modes [Fig. 3(b)] it seems probable that the frequency difference between the two $c/2L$ components should increase with the difference in frequency between the central cavity resonance and the center of the Doppler line. Hence, for example, the two $c/2L$ beats could be used to generate an error frequency (whose sharpness would be determined by the linewidth of the oscillation)[2,11] which could be used in a magnetostrictive tuning device to stabilize the central mode near the center of the Doppler line. The second consequence follows from a reasonably general symmetry argument. If we assume that the Doppler line shape is perfectly symmetric about its center frequency and that there are only three modes oscillating, any splitting in the $c/2L$ beats must arise from an asymmetric distribution of the three modes about the Doppler center. That is, if the central cavity mode is precisely on the Doppler center, the pulling on the other two resonances must be equal and opposite; hence in this case no splitting can occur.

4. Parameters in the Present Maser

For convenience, we summarize the pertinent parameters characteristic of the present maser in this section.

The frequency separation of the cavity resonances ($c/2L$) is nominally $\cong 160$ Mc/sec (Table 1). The wavelength of the maser

[10] A. J. Rack (private communication).

[11] The linewidths implied by the data in Fig. 3 are limited by the resolution of the spectrum analyzer.

transition is 11,522.76 Å and corresponds to the $2s_2$ to $2p_4$ line of neon in Paschen notation.[3] From direct gain measurements on this transition under similar discharge conditions, the fractional energy gain in a single pass through the interferometer should be about 6%. The known reflectance losses[4] ($\cong 1\%$) imply a cavity width ($\Delta\nu_c$) of about $\frac{1}{2}$ Mc/sec [Eq. (3) below]. However, the data in Table 1 taken with the analysis below, demonstrate that $\Delta\nu_c$ is closer to 1 Mc/sec and that the total energy loss per pass is about 2%. Hence the maximum gain available corresponds to about three times the threshold for oscillation. The latter is in good agreement with the maximum number of modes observed [Fig. 3(d)] and the 800 Mc/sec Doppler width assumed below.

In the limit of complete resonance trapping of vacuum ultraviolet photons, the effective decay rates of the upper and lower maser levels are about 10^7 sec^{-1} and 8×10^7 sec^{-1}, respectively.[12] The exact value of the natural linewidth ($\Delta\nu_n$) for the maser transition depends critically on the unknown, vacuum ultraviolet decay rate of the upper maser level. It is expected, however, that $15 < \Delta\nu_n < 80$ Mc/sec (see Sec. 10).

The full Doppler width at half-maximum intensity for neon atoms at the temperature of the maser is about 800 Mc/sec. The inelastic collision with the He($2\,^3S$) atom by which the excited neon state is formed[1] is, of course, exothermic by $\cong 1.5kT$ at room temperature. However, since the He atom has $\frac{1}{5}$ the mass of the neon atom, it will carry off most of the energy by which the reaction is exothermic. Although the Doppler width will increase with the gas temperature and hence with the power output in the maser, it does so as the square root of the absolute temperature. Hence a Doppler width much in excess of 800 Mc/sec is not be be expected. It is inconceivable, for example, that the Doppler width could increase sufficiently with power to explain the magnitude of the anomalous power dependence of the beat frequencies reported above.

[12] W. R. Bennett, Jr., in *Advances in Quantum Electronics*, edited by J. R. Singer (Columbia University Press, New York, 1961), p. 28.

5. Mode Pulling Analysis

We are only concerned with the dominant modes of even radial symmetry corresponding to longitudinal propagation in a plane parallel Fabry-Perot interferometer. In what follows it is assumed that steady-state oscillation has been obtained in the system and parallel Febry–Perot interferometer. In what follows it is assumed that no coupling effects exist between simultaneously oscillating modes through time-dependent nonlinearities in the medium. These assumptions are likely to break down at some point. However, what is sought are manageable, approximate solutions which well explain the dominant mode pulling effects.

The oscillation frequency in the maser is determined primarily by a condition on the phase of the electric field and the phase shifts of importance all arise from single-pass time delays. By definition, the phase shift resulting from a wave traveling once through an interferometer of length L at a phase velocity c/n (where n is the refractive index) is

$$\phi = 2\pi \nu Ln/c. \tag{1}$$

For a standing wave to build up in the cavity, the single-pass phase shift must be an integral multiple of π. Hence the evacuated cavity ($n = 1$) has resonant frequencies which are separated by $c/2L$.

From Eq. (1), the dispersion ($\partial\phi_c/\partial\nu$) for the evacuated cavity ($n = 1$) is a constant. It is convenient to express $\partial\phi_c/\partial\nu$ in terms of the fractional energy loss per pass f and the full width of the cavity resonance at half-maximum intensity, $\Delta\nu_{0c}$. To accomplish the latter, we visualize a situation in which energy (U) is placed in the evacuated cavity in the mode of interest. This energy decays exponentially with time at the rate $(c/L)f$. As a result, the frequency response of the interferometer is not perfectly sharp and one may define a Q for the evacuated cavity given by

$$Q_{0c} = \frac{2\pi\nu_{0c}U}{(cf/L)U} = \frac{\nu_{0c}}{\Delta\nu_{0c}}. \tag{2}$$

From (1) and (2),

$$\partial\phi_c/\partial\nu = 2\pi L/c = f/\Delta\nu_{0c}$$
$$= \text{const} \cong 2 \times 10^{-8} \text{ rad-sec.} \tag{3}$$

For the present system, a single-pass loss of one percent corresponds to $\Delta\nu_{0c} \cong (1/2)$ Mc/sec.

There will, in general, be some entirely negligible contribution to $(\partial\phi_c/\partial\nu)$ arising from the resonant nature of the mirror reflectance coefficient: For example, in the present case the mirror transmission loss varied by a factor of two over a total range of about 1000 Å (or 2×10^{12} cycles/sec). If one then estimates the dispersion at maximum reflectance using a Lorentzian line shape [Eq. (9)], it is seen that the contribution from the mirrors is about 5×10^{-15} rad-sec. Hence the inclusion of this term alters Eq. (3) by about 2 parts in 10^7. The inclusion of this term in the analysis given below merely changes the final pulling term in Eq. (22) by the same fractional amount and has no important effect on the absolute frequency of the oscillation. We therefore ignore this contribution to $(\partial\phi_c/\partial\nu)$.

The introduction of the amplifying medium changes the refractive index in the system, thereby altering the single-pass phase shift from that obtained in the evacuated case. Oscillation therefore occurs at another frequency $\bar{\nu}$, differing from the cavity resonance frequency ν_{0c}, such that the single-pass phase shift is still an integral multiple of π. Since the cavity dispersion is large compared to that for the amplifying medium, oscillation occurs close to ν_{0c} and the pulling is small. It is convenient to formulate the problem in terms of the difference in frequency from the cavity resonance. In particular, the maser oscillates at that frequency $\bar{\nu}$ such that

$$(\partial\phi_c/\partial\nu)(\bar{\nu} - \nu_{0c}) + \Delta\Phi_m(\bar{\nu}) = 0\,. \tag{4}$$

Here, $\Delta\Phi_m(\bar{\nu})$ is the total change in single-pass phase shift at the actual frequency of oscillation which is caused by the insertion of the medium. $\Delta\Phi_m(\bar{\nu})$ is composed of two parts and, from Eq. (1), may be expressed as

$$\begin{aligned} \Delta\Phi_m(\bar{\nu}) &= (2\pi L/c)[(n_0 - 1) + (n - 1)]\bar{\nu} \\ &\equiv (2\pi L/c)(n_0 - 1)\bar{\nu} + \Delta\phi_m(\nu)\,. \end{aligned} \tag{5}$$

The first term in Eq. (5), $(2\pi L/c)(n_0 - 1)\bar{\nu}$, arises from the density of ground-state atoms in the maser and from the density of excited atoms which may participate in neighboring transitions (e.g., the $2\,^3S - 2\,^3P$ of He at 10,830 Å). That is, this first term arises from a refractive index which is essentially independent of the frequency over the range of interest. Its main source in the helium–neon maser is the He(^1S) at 1 mm Hg for which $(n_0 - 1) \cong 5 \times 10^{-8}$.[13]

From Eqs. (3) and (5), Eq. (4) may be expressed,

$$\bar{\nu} = \left(\frac{\nu_{0c}}{n_0}\right) - \left(\frac{\Delta\nu_{0c}/n_0}{f}\right)\Delta\phi_m(\bar{\nu}).$$

We include the effects of n_0 by defining

$$\bar{\nu}_c = \nu_{0c}/n_0 \quad \text{and} \quad \Delta\nu_c = \Delta\nu_{0c}/n_0. \tag{6}$$

Hence the oscillator frequency ($\bar{\nu}$) is given by

$$\bar{\nu} = \nu_c - (\Delta\nu_c/f)\,\Delta\phi_m(\bar{\nu}). \tag{7}$$

For the helium–neon case, $\Delta\nu_c$ in Eq. (6) differs negligibly ($\cong 5$ parts in 10^8) from $\Delta\nu_{0c}$ in Eq. (3). Similarly, the frequency separation between adjacent cavity resonances differs by the same negligible constant amount from $c/2L$. For an arbitrary setting of the interferometer, n_0 does introduce a pressure-dependent shift in the oscillator frequency of about 13 Mc/sec per mm Hg of He. However, in the special case that ν_c is tuned to the center of a symmetric line (see below), the oscillator frequency is independent of ν_c and this pressure-dependent shift vanishes.

The term $\Delta\phi_m(\bar{\nu})$ in Eqs. (5) and (7) is a function of the fractional energy gain per pass $g(\bar{\nu})$. Since the latter varies with frequency over the transition responsible for the amplification $\Delta\phi_m$ does also. Generally, $\Delta\phi_m$ goes through zero at the line center (ν_m), is

[13] *American Institute of Physics Handbook* (McGraw-Hill Book Company, Inc., New York, 1957), pp. 6–21.

negative ("anomalous dispersion") for frequencies less than ν_m, and is positive for frequencies greater than ν_m. Equation (6) therefore predicts a shift in the direction to the line center.

Threshold for oscillation occurs at that frequency $\bar{\nu}_T$, satisfying both Eq. (7) and the condition $g(\bar{\nu}_T) = f$. The latter generally results in a transcendental equation for $\bar{\nu}_T$. In the special case of a Lorentzian line, the transcendental nature of this equation is removed.

The existence of a steady-state oscillation above threshold obviously implies that the gain at the frequency of oscillation must saturate at

$$g(\bar{\nu}) = f. \tag{8}$$

In the case of homogeneous or "natural" broadening, gain proportionality is always maintained over the line. That is, the reduction of the gain at frequency $\bar{\nu}$ necessary to satisfy Eq. (8) produces a proportionate reduction of the gain at all other frequencies. Hence in the case of homogeneous broadening, the oscillation frequency is always given by its value at threshold and there is no direct power-dependent pulling effect. In the case of inhomogeneous broadening (applying to the He–Ne maser), the interpretation of requirement (8) is more complicated.

6. Assumptions on Line Shape

In general, $\Delta\phi_m$ will be a function of the single-pass gain and for a given line shape could be calculated numerically from the Kramers–Kronig relations.[14]

For a Lorentzian line, it may be shown (see Appendix I) that

$$\Delta\phi_m(\nu) \cong -\frac{g_m(\nu_m - \nu)}{\Delta\nu_m}\left[1 + \frac{4(\nu_m - \nu)^2}{(\Delta\nu_m)^2}\right]^{-1}, \tag{9}$$

[14] D. E. Thomas (to be published) has extended his "Tables of Phase of a Semi-Infinite Unit Attenuation Slope" (Bell System Monograph 2550) to handle general problems of the present type numerically. His results for the special case of a Gaussian have been verified analytically by A. Javan.

where the fractional energy gain per pass at the frequency ν is given by

$$g(\nu) \cong g_m[1 + 4(\nu_m - \nu)^2/(\Delta\nu_m)^2]^{-1}. \tag{10}$$

Here, as in the rest of this paper, we adopt the notation that g_m is the fractional energy gain per pass at the time center (ν_m) and that $\Delta\nu_m$ is the full width of the line at half-maximum fractional energy gain. The approximations in (9) and (10) arise from the assumption that $\nu + \nu_m = 2\nu$ and that $g_m \ll 1$. Equations (9) and (10) may be expressed

$$\Delta\phi_m(\nu) \cong -g(\nu)(\nu_m - \nu)/\Delta\nu_m. \tag{11}$$

The Lorentzian form will be used in this paper to obtain an accurate expression for the pulling in the case of natural broadening and to allow for the holes produced by inhomogeneous broadening in the case of the helium–neon maser.

What actually is involved in the helium–neon maser is a Gaussian distribution of Lorentzian lines resulting from the thermal motion of the excited neon atoms and the corresponding Doppler shift of the atoms' center frequencies. Hence, in the limit (holding here) that the natural linewidth is small compared to the Doppler width, the line much more closely resembles a Gaussian than a Lorentzian.

We are primarily concerned with effects taking place within the full width of the line at half-maximum intensity and in the limit of small energy gain. In the limit of small gain, it is apparent from the Kramers–Kronig relations[15] that the phase shift increases linearly with the gain. Hence for any typical symmetric line, the phase shift introduced by the amplifying medium may be expanded in the series

$$\Delta\phi_m(\nu) \cong -ag_m\left(\frac{\nu_m - \nu}{\Delta\nu_m}\right)\left[1 - b\left(\frac{\nu_m - \nu}{\Delta\nu_m}\right)^2 + \cdots\right], \tag{12}$$

where the constants a and b will depend on the line shape. Since $\Delta\phi_m$ must have odd symmetry about the line center, approximation

[15] See, for example, J. H. Van Vleck, *Radiation Laboratory Series* (McGraw-Hill Book Company, Inc., New York, 1948), Vol. 13, Chap 8.

(12) involves the assumption that $[(\nu_m - \nu)/\Delta\nu_m]^4 \ll 1$. Hence, approximation (12) will be extremely good over a considerable range within the central portion of the line and will fail rapidly for frequencies occurring near the half-intensity points on the gain curve. Approximation (12), of course, consists of the first two terms in the expansion of the sine functions and for manipulative reasons it is convenient to express $\Delta\phi_m$ in the latter way. In particular, a "best fit" of numerical calculations made by Thomas[14] demonstrates that

$$\Delta\phi_m(\nu) \cong -0.28_2 g_m \sin\left(\frac{\nu_m - \nu}{0.3\Delta\nu_m}\right) \tag{13}$$

is an exceptionally good approximation for a Gaussian. Both "angles" in Eq. (13) are expressed in radians. The errors introduced by a literal interpretation of (13) are less than 1% for $|\nu_m - \nu| \leq 0.4\Delta\nu_m$. Equation (13) is $\cong$6% low at the half-intensity points on the curve. For $|\nu_m - \nu| > 0.5\Delta\nu_m$ the error increases exponentially. The corresponding fractional energy gain for the Gaussian is,

$$g(\nu) \cong g_m \exp\left[-\left(\frac{\nu_m - \nu}{0.6\Delta\nu_m}\right)^2\right]. \tag{14}$$

From (13) and (14)

$$\Delta\phi_m(\nu) \cong -0.28_2 g(\nu) \exp\left[\left(\frac{\nu_m - \nu}{0.6\Delta\nu_m}\right)^2\right] \sin\left(\frac{\nu_m - \nu}{0.3\Delta\nu_m}\right) \tag{15}$$

holds for a Gaussian within the central region of the curve.

7. Homogeneous Broadening

For a homogeneously broadened Lorentzian line, (7), (8), and (11) yield

$$\bar{\nu} = (\nu_c\Delta\nu_m + \nu_m\Delta\nu_c)/(\Delta\nu_m + \Delta\nu_c) \tag{16}$$

for the oscillation frequency where definition (6) is assumed.[16] In

[16] Equation (16) is equivalent to a result quoted by C. H. Townes, in *Advances in Quantum Electronics*, edited by J. R. Singer (Columbia University Press, New York, 1961), p. 10.

the limit that $\Delta\nu_c < \Delta\nu_m$, (16) becomes

$$\bar{\nu} = \nu_c + (\nu_m - \nu_c)\left(\frac{\Delta\nu_c}{\Delta\nu_m}\right)\left[1 - \frac{\Delta\nu_c}{\Delta\nu_m} + \cdots\right], \tag{17}$$

and in the Lorentzian case, the pulling is linearly dependent on the frequencies. The latter would, of course, not be true in the case of an inhomogeneously broadened Lorentzian line above threshold. As may be seen by comparing the expanded terms of Eqs. (9) and (13), the phase characteristic of the Lorentzian is actually about twice as nonlinear as that of the Gaussian near the line center.

For a Gaussian line, (7), (8) and (15) yield,

$$\bar{\nu} = \nu_c + 0.28_2\Delta\nu_c \exp\left[\left(\frac{\nu_m - \bar{\nu}}{0.6\Delta\nu_m}\right)^2\right] \sin\left(\frac{\nu_m - \bar{\nu}}{0.3\Delta\nu_m}\right), \tag{18}$$

and the pulling is nonlinear even in the homogeneously broadened case. For frequencies very near the line center in the limit that $\Delta\nu_c < \Delta\nu_m$, Eq. (18) becomes

$$\bar{\nu} = \nu_c + (\nu_m - \nu_c)(0.94\,\Delta\nu_c/\Delta\nu_m) \times [1 - 0.94(\Delta\nu_c/\Delta\nu_m) + \cdots]. \tag{19}$$

Hence, for frequencies near the center of the line in the limit $\Delta\nu_c \ll \Delta\nu_m$, the main difference between the Gaussian and the Lorentzian is a 6% reduction in the pulling factor.

Equation (18) only applies to the helium–neon maser in the case of threshold for the first cavity resonance. For an inhomogeneously broadened line, an expansion of the threshold pulling terms in powers of $\Delta\nu_c/\Delta\nu_m$ is misleading. The neglected nonlinear terms are generally much larger than $(\Delta\nu_c/\Delta\nu_m)^2$. The latter is apparent for Eq. (18), but is also true of Eq. (16). The inhomogeneously broadened case may be more realistically treated through the method given below.

8. General Approximate Solution for Optical Masers

Generally for optical masers, $\Delta\nu_m \gg \Delta\nu_c$ and $\Delta\phi_m(\nu)$ will be a slowly varying function of the frequency over the cavity resonance.

We may therefore expand $\Delta\phi_m$ in a Taylor series about the cavity resonance frequency, obtaining

$$\Delta\phi_m = (\Delta\phi_m)_{\nu_c} + (\partial\Delta\phi_m/\partial\nu)_{\nu_c}(\nu - \nu_c) + \cdots. \tag{20}$$

Substituting Eq. (20) in Eq. (7) yields

$$\bar{\nu} = \nu_c - (\Delta\nu_c/f)\Delta\phi_m(\nu_c) \times [1 - (\Delta\nu_c/f)(\partial\Delta\phi_m/\partial\nu)_{\nu_c} + \cdots]. \tag{21}$$

As may be seen by comparison with Eqs. (17) and (19), this last approximation is equivalent to an expansion of the pulling terms in powers of $(\Delta\nu_c/\Delta\nu_m)$ which still retains the nonlinear properties of the phase characteristic. Typically, $(\Delta\nu_c/\Delta\nu_m) \cong 10^{-3}$. Hence the oscillator frequency is given by

$$\bar{\nu} \cong \nu_c - (\Delta\nu_c/f)\Delta\phi_m(\nu_c), \tag{22}$$

where errors in the second term of about a part in 10^3 (i.e., about 200 cycles/sec in the absolute frequency for the helium–neon maser) may be expected. $\Delta\phi_m(\nu_c)$ represents the actual phase shift introduced by the amplifying transition at the cavity resonance in the presence of oscillation. Equation (22) may be solved numerically using the methods of Thomas[14] in the general case (e.g., asymmetric lines) for both homogeneous and inhomogeneous broadening.

9. Effect of Holes

It is in the interpretation of requirement (8) that an important difference arises between the helium–neon maser and, for example, the ammonia maser (or more generally, between the present oscillator and most other oscillators). In the present system, the line is primarily broadened by the Doppler effect. Consequently, reduction of the gain at the frequency $\bar{\nu}$ in order to satisfy Eq. (8) does not imply a proportionate reduction of the gain at other frequencies. If the converse were true (as in the case of homogeneous broadening), a second cavity mode would generally not go into oscillation; i.e., the saturation requirement (8) at the first resonance would prevent the second cavity resonance from reaching threshold. In the present case, (8) is satisfied by burning a hole in the line. $g(\bar{\nu})$ saturates at f, whereas the gain over the rest of the line continues to increase with pumping rate. In other words, the phase shift introduced at the cavity resonance continues to increase as the number of upper state atoms increases at frequencies well removed from the hole. We shall satisfy requirement (8) in the following analysis by subtracting the phase shift which would have been produced by atoms in the hole from the phase shift introduced by the entire distribution in the absence of oscillation. In evaluating these phase shifts, we make use of the same approximation implicit in Eq. (22)—namely, that $\bar{\nu}$ may be replaced by ν_c.

Before taking explicit account of the holes, we make the following two observations:

(a) Since atoms on opposite sides of the cavity resonance (ν_c) contribute to the gain at ν_c with the same sign, the total gain at ν_c is determined primarily by those atoms whose center frequencies fall within the resonance by about one natural linewidth. Hence, variations in the upper state density at large frequency separations from the cavity resonance do not affect the gain at ν_c appreciably, and a hole burned at one resonance does not have an important first-order effect on the gain at another resonance.

(b) Since atoms on opposing sides of the cavity resonance contribute to the phase shift at ν_c with opposite sign, the net phase shift at ν_c is determined primarily by atoms which are well removed from the cavity resonance. Hence the absence of atoms (i.e., a hole) at one resonance can have a large first-order effect on the phase shift at another resonance. The presence of a hole at one resonance reduces the pulling at another resonance which would have existed in the absence of the hole. That is, two holes always tend to repel each other. However, since a hole represents a symmetric removal of atoms about a cavity resonance to first order, a hole does not have a first-order effect on itself.

It is shown in Appendix II that the probability for stimulated emission is a Lorentzian function of the frequency separation between the mode of oscillation and the center frequency of the atoms involved. In addition, it is shown that the full width at half-maximum for this probability distribution is a function of the power in the mode and, hence, of the gain at the resonance. We therefore assume that the holes burned in the line will be Lorentzian in shape and have widths dependent on their location in respect to the line center. Although the Lorentzian hole shape would only hold strictly for a constant distribution of atoms, it is apparent that it will represent a good first-order approximation at low powers in the present case.

By analogy with Eq. (9), the phase shift introduced at frequency ν_2 by a Lorentzian hole centered at frequency ν_1, is

$$\Delta\phi_H(\nu_2) = \frac{D_1(\nu_1 - \nu_2)}{H_1}\left[1 + \frac{4(\nu_1 - \nu_2)^2}{{H_1}^2}\right]^{-1}, \tag{23}$$

where D_1 is the depth of the hole at ν_1 and H_1 is the full width of this hole at half-maximum intensity. The sign on Eq. (23) is opposite to that for Eq. (9) since we are concerned with phase shifts introduced by the absence of atoms. Since we are only interested in the phase shift produced at one resonance by holes at different resonances [observation (b) above], $|\nu_1 - \nu_2| \geqq (c/2L)$. In the low power limit, we may assume that the hole widths are less than the hole separations and consequently that $4(\nu_1 - \nu_2)^2 \gg {H_1}^2$. This approximation introduces an error of about 5% in the evaluation of $\Delta\phi_H$ for the hole widths required below to explain the anomalous increase in beat frequency with pumping rate. Therefore,

$$\Delta\phi_H(\nu_2) \cong \left(\frac{g(1) - f}{4}\right)\left(\frac{H_1}{\nu_1 - \nu_2}\right), \tag{24}$$

where we have defined $D_1 = g(1) - f$ in conformance with saturation condition (8). This choice of hole depth is illustrated in Fig. 4(a) for the case of three resonances.

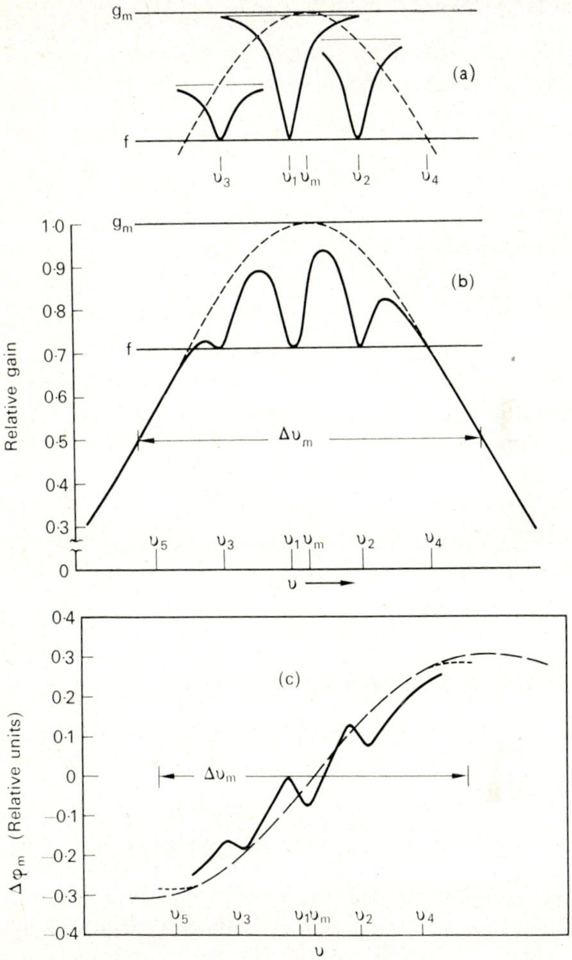

FIG. 4. (a) Illustrates the choice of hole depth used to satisfy Eq. (8). (b) illustrates the resultant line shape at threshold for the fourth cavity resonance when $\nu_m - \nu_1 = 40$ Mc/sec. $\Delta\nu_m = 800$ Mc/sec and $c/2L = 160$ Mc/sec. Equal hole widths of 64 Mc/sec have arbitrarily been assumed. (c) (the solid curve) illustrates the resultant variation of the total phase shift for the condition in (b). The dashed curve represents the phase shift introduced by the original Gaussian. The dotted curve shows the departure of approximation (13) from the actual phase characteristic of the Gaussian.

The total phase shift introduced at frequency ν_j by N Lorentzian holes at frequencies ν_i is then

$$\Delta\phi_H(j) \cong \sum_{i=1(i\neq j)}^{N} \left(\frac{g(i)-f}{4}\right)\left(\frac{H_i}{\nu_i - \nu_j}\right). \tag{25}$$

From Eqs. (22) and (25) the oscillation frequency for the jth cavity resonance is given by

$$\bar{\nu}_j \cong \nu_j - \frac{\Delta\nu_c}{f}\,\Delta\phi_m(\nu_j) - \frac{\Delta\nu_c}{4}\sum_{i=1(i\neq j)}^{N}\left(\frac{g(i)}{f} - 1\right)\left(\frac{H_i}{\nu_i - \nu_j}\right), \tag{26}$$

where by definition $g(i) \geqq f$ for a hole to occur at frequency ν_i. The quantities $g(i)$ are related by Eq. (14) and $\Delta\phi_m$ is given by Eq. (13) for a Gaussian line. For a Lorentzian line, Eqs. (9) and (10) obtain.

A numerical evaluation of the total phase shift over the central portion of a Gaussian line is illustrated in Fig. 4(c) for three Lorentzian lines. The errors introduced by approximation (13) are also indicated in Fig. 4(c).

10. Hole Widths from Stimulated Emission

From Eq. (II.11) of Appendix II, the full width (H) of a hole produced by stimulated emission is

$$H = \left(\frac{\gamma_a + \gamma_b}{2\pi}\right)\left(1 + \frac{1}{\gamma_a\gamma_b}\left|\frac{2V}{\hbar}\right|^2\right)^{\frac{1}{2}}, \tag{27}$$

γ_a and γ_b represent the decay rates of the upper and lower maser levels, respectively, and the matrix element V is given by

$$|V|^2 = |eE_0(z)_{a,b}/2|^2, \tag{28}$$

where E_0 is the amplitude of the electric field in the cavity mode. E_0^2 is proportional to the power (P) expended in the mode and may be readily expressed the latter way through the known cavity Q and volume ($v = 108\ \text{cm}^3$) of the discharge;

$$E_0^2 = 4P/(v\Delta\nu_c). \tag{29}$$

The upper maser level may decay by several transitions, hence

$$\gamma_a = \gamma_{ab} + \gamma_{ac} + \cdots, \tag{30}$$

where

$$\gamma_{ab} = \frac{4}{3}\frac{e^2}{\hbar\lambda^3}|r_{ab}|^2 \cong \frac{4e^2}{\hbar\lambda^3}|z_{ab}|^2, \tag{31}$$

where λ and $\hbar$ are, respectively, the wavelength for the transition and Planck's constant divided by 2π. Combining Eqs. (27), (28), (29), and (31) yields

$$H \cong \frac{\gamma_a + \gamma_b}{2\pi}\left(1 + \frac{\gamma_{ab}}{\gamma_a\gamma_b}\frac{\lambda^3 P}{\hbar v \Delta\nu_c}\right)^{\frac{1}{2}}, \tag{32}$$

where cgs units are used and the coefficient outside the square root,

$$\Delta\nu_n = (\gamma_a + \gamma_b)/2\pi, \tag{33}$$

represents the natural linewidth for the transition.

If the decay rates in Eq. (32) were determined entirely by known spontaneous radiative values, the interpretation of Eq. (32) would be straightforward. The effective decay rates for the two maser levels have approximate values[12] of

$$\begin{aligned} \gamma_a &\cong 10^7 \text{ sec}^{-1}, \\ \gamma_b &\cong 8 \times 10^7 \text{ sec}^{-1}. \end{aligned} \tag{34}$$

However, the vacuum ultraviolet portion of the decay rate of the upper maser-level is unknown. The resonance trapping process [which determines the effective, long lifetime of the upper maser level given in Eq. (34)] will interrupt the phase of the wave function used to obtain Eq. (27) by some unknown, large extent. Consequently, an unambiguous interpretation of Eq. (32) cannot be given and we may only examine two limiting cases. The first limit is obtained by inserting the values (34) in Eq. (32). Here one obtains the minimum hole width (15 Mc/sec) at threshold and the maximum dependence of the hole width on power. From the measured power in the beam and estimates of the other parameters involved, the hole widths in this limiting case might extend to about 50 Mc/sec at the highest output powers obtained. The second limit is obtained through the marginal observation of a clean $c/2L$ beat at low power in another He–Ne maser having twice the length of the present one. This observation implies that the natural linewidth for the transition is less than 80 Mc/sec. Hence in the second limiting case maximum hole width of about 100 Mc/sec might be obtained at high power from stimulated emission.

A related point of interest arises here which should be noted. Namely, because of the phase-interruption process, more output power is to be expected from a maser transition in which the upper state is optically connected with the ground state than for one which is not. That is, the hole widths in the former case will be larger, even if the effective decay rates and oscillator strengths for the two maser transitions are identical.

11. Effects of Elastic Scattering

There exist only two likely sources by which the upper state atoms may migrate over the Doppler distribution during their lifetimes:

(1) large-angle elastic scattering.
(2) small-angle elastic scattering.

We have separated the elastic (atom-atom) scattering process into these two groups both because the nature of the differential scattering cross section readily permits this separation and because the effects of the two processes are in opposite directions.

Process (1) results in violent changes in velocity of the upper state maser atoms and, hence, in changes in the location of the atoms' center frequencies which are comparable to the full width of the Doppler line. This effect therefore tends to restore the gain proportionality condition which holds, for example, over the entire line in the case of natural broadening. Thus, process (1) tends to oppose the burning of holes in the line. Its effects, however, are small in the present case. The total large-angle scattering cross section for atoms in the upper neon maser state on ground state He atoms is unknown. However, the known[17] diffusion rates of the metastable He($2\,^3S_1$) and Ne($3\,^3P_2$) levels in their own gases both correspond to total ("hard sphere") large-angle scattering cross sections of about 10^{-15} cm^2 at a pressure of 1 mm Hg, and a temperature of 300°K. Since the upper Ne maser level has a similar configuration ($4\,^1P_1$ is LS notation) to the Ne($3\,^3P_2$), it seems unlikely that the total large-angle elastic scattering cross section would be appreciably different in the present case. Hence we expect the total large-angle collision rate for the upper maser level for the pressures in the maser (1 mm Hg of He and 0.1 mm Hg of Ne) to be about 0.5×10^7 sec^{-1}. The latter corresponds to about half the known[12] decay rate of the upper maser level.

Process (1) therefore takes place $\cong 0.5$ times per atom on the average and the probability of the atom landing in a hole at low powers may be neglected.

The small-angle collisions, however, will result in the widening of a hole burned in the line. In the real case, the tail of the Van der Waals interaction, the possibility of molecular bonds, etc., will increase the differential scattering cross section enormously at small angles. The magnitudes of these interactions are unknown in the present case. However, one may estimate the size of the effect by assuming a hard-sphere cross section corresponding to typical diffusion rates. From the quantum mechanical treatment of the hard-sphere problem,[18] there is always a cone of small, finite, half-angle

$$\theta \cong h/2\mu_m v_{\mathrm{rel}} a\,, \tag{35}$$

within which the scattering is nonclassical and within which the total scattering rate is generally about equal to the total large-angle scattering rate. Here, θ is given in the center of mass system, μ_m is the reduced mass, v_{rel} the relative velocity, h is Planck's constant, and a is the molecular diameter. This type of effect is difficult to observe experimentally and would not be included, for example,

[17] A. V. Phelps and J. P. Molnar, *Phys. Rev.* **89,** 1202 (1953).

[18] N. F. Mott and H. S. W. Massey, *The Theory of Atomic Collisions* (Clarendon Press, Oxford, England, 1949), 2nd ed., Chap. II.

P*

in the elastic scattering cross section measured in a diffusion experiment. We next assume that the average, small-angle scattering event corresponds to scattering in the center-of-mass system through an angle $\theta/2$. This average small-angle collision between the excited neon atom and a ground-state helium atom therefore causes the excited neon atom to change its center frequency by an amount

$$\delta\nu = \delta v_{\mathrm{Ne}}\nu/c = \delta v_{\mathrm{rel}}\nu/6c \cong h\nu/24\mu_m ca\,. \tag{36}$$

We next choose "a" to correspond to the classical hard-sphere scattering cross section determined from the diffusion data in reference 17 (i.e., $\pi a^2 = 10^{-15}$ cm^2) and obtain the result

$$\delta\nu \cong 30\ Mc/\mathrm{sec}. \tag{37}$$

That is, the typical step is about equal to 30 Mc/sec and on the average half the atoms will make one such step in their lifetimes. Hence holes burned in the Doppler line might be enlarged by $\cong$15 Mc/sec due to small-angle scattering.

12. Semi-empirical Representation of Hole Widths

From the discussion in Secs. 10 and 11, the hole widths at low powers will approach an unknown constant value falling in the tens of Mc/sec range. From the discussion in Sec. 10, it is apparent that well above threshold the hole widths will increase as the square root of the power in the mode. The power in the mode itself will be approximately proportional to the product of the hole width and the hole depth. Hence at high powers one expects

$$H_i \cong H_0[g(i) - f]/f,$$

where the constant H_0 will also be in the tens of Mc/sec range. Hence we may represent the width of the ith hole by

$$H_i \cong h_0 + H_0[g(i) - f]/f \tag{38}$$

over the entire range in power. The adoption of Eq. (38) permits the evaluation of Eq. (26) with a minimum number of adjustable parameters.

13. Comparison of Mode Pulling Analysis with Experiment

Since the distribution of atoms is peaked at frequency ν_m, there will be a definite order of appearance of the various possible cavity resonances. We shall enumerate these cavity resonance frequencies in order of their appearance:

$$\nu_1, \nu_2, \nu_3 \cdots, \quad \text{where} \quad \cdots \nu_3 < \nu_1 < \nu_m < \nu_2 \cdots. \tag{39}$$

Condition (39) corresponds to the one illustrated in Fig. 4 and will be assumed below in the cases for which Eq. (26) is evaluated. In making the present comparisons we consider only the Gaussian line for which Eqs. (13), (14), and (15) apply.

Evaluation of Eq. (26) for the first two resonances ($\nu_2 > \nu_1$) yields

$$\bar{\nu}_2 - \bar{\nu}_1 \cong \left(\frac{c}{2L}\right) + \frac{\Delta\nu_c}{4(c/2L)}\left[\left(\frac{g_1 - f}{f}\right)H_1 + \left(\frac{g_2 - f}{f}\right)H_2\right]$$
$$- 0.56_4 \frac{g_m}{f}\Delta\nu_c \sin\left(\frac{c/2L}{0.6\Delta\nu_m}\right)\cos\left(\frac{\nu_m - \nu_1 - c/4L}{0.3\Delta\nu_m}\right), \tag{40}$$

where $g(1)$, $g(2)$, and g_m are related through Eq. (14) and it is assumed that $g(2) \geqq f > g(3)$.

The first case of interest corresponds to threshold for the appearance of the $c/2L$ beat and, therefore, the data in Table 1. Here, Eq. (40) reduces to

$$\bar{\nu}_2 - \bar{\nu}_1 \cong (c/2L)[1 - 0.94(\Delta\nu_c/\Delta\nu_m)] \tag{41}$$

plus terms of order $\Delta\nu_c(c/2L\Delta\nu_m)^3$. The latter amount to frequency fluctuations in the order of 10 kc/sec and are within the limits of error quoted in Table 1. Comparison of Eq. (41) with the data in Table 1 implies that $\Delta\nu_c \cong 1$ Mc/sec, since $\Delta\nu_m \cong 800$ Mc/sec. This value of $\Delta\nu_c$ is in discrepancy with the value ($\cong$0.5 Mc/sec) obtained from Eq. (3) for the known (0.99) mirror reflectance coefficient and suggests that some additional loss of about 1% per pass is present in the maser other than that introduced by the mirror reflectance coefficient. Fox and Li[19] have made tentative estimates which indicate that a loss of this approximate magnitude might arise from mode mixing at the end plates from flatness irregularities. In support of this contention we note that the present maser obviously has more loss than the first one reported[1] since it oscillated only on the strongest of the five transitions of neon previously observed. Here the main difference between the two masers is that the first maser had Fabry–Perot plates of considerably higher quality. It is conceivable that the reflectance films in the present maser deteriorated by the required amount during bakeout,[8] in which case the contention is not clearly established. It was found, however, that the loss implied by the data in Table 1 remained constant within the limits of error quoted over six months of continuous operation. In what follows, we assume that the ratio $\Delta\nu_c/\Delta\nu_m \cong 1/800$ has been experimentally determined from the data in Table 1.

The second term in Eq. (40) increases with the power and consequently is capable of explaining the anomalous behavior of the $c/2L$ beat reported in Sec. 3. The hole repulsion effect, however, can only explain this anomalous behavior if it dominates over the third term in Eq. (40). The variation of the beat frequency above threshold will obviously be a complicated function of the location of ν_1 in respect to the line center. There are two extreme limiting cases and we only consider these. Because of the large number of exponential terms, it is simplest to solve Eq. (40) numerically. The results quoted below are for the parameters characteristic of the present maser—namely $c/2L \cong 160$ Mc/sec, $\Delta\nu_m \cong 800$ Mc/sec, and $\Delta\nu_c \cong 1$ Mc/sec. In evaluating Eq. (40) we express the total excursion of the beat between threshold for its appearance and threshold for the next cavity resonance. Hence, through the use of Eqs. (14) and (38), only two adjustable parameters h_0 and H_0 appear in the final result.

The first limiting case corresponds to a symmetric placement of the two cavity resonances about the Doppler line. Here the minimum hole repulsion effect should occur since both resonances cross

[19] A. G. Fox and Tingye Li (private communication).

threshold simultaneously and the two hole widths must be the same minimum value at threshold. A numerical solution of Eqs. (14), (38), and (40) for the present maser in this limiting case yields

$$(\bar{\nu}_2 - \bar{\nu}_1)_{\text{max}} - (\bar{\nu}_2 - \bar{\nu}_1)_{\text{thresh}} \cong 0.75[h_0 + 0.25H_0 - 60] \text{ kc/sec}, \tag{42}$$

where the bracketed quantities are expressed in Mc/sec. The fact that an interferometer setting could be found (Sec. 3) for which the $c/2L$ beat was nearly independent of power suggests that the bracketed quantity in Eq. (42) is very close to zero. Hence we expect

$$h_0 + 0.25H_0 \cong 60 \text{ } Mc/\text{sec} \tag{43}$$

for the present maser.

The other limiting case occurs at $\nu_1 = \nu_m$. Here the first hole may be enlarged considerably through stimulated emission before the second resonance occurs and the maximum hole repulsion effect should arise. This limit is complicated in the real case by the fact that both the second and third cavity resonances cross threshold simultaneously. Here the two $c/2L$ beats are identical and splitting (see below) does not occur until the fourth and fifth cavity resonances cross threshold. The effect of the third hole will not alter the present estimate substantially, however, and we shall ignore its presence (i.e., the effects of the holes at ν_2 and ν_3 cancel at ν_1, whereas the effects of the holes at ν_1 and ν_3 are additive at ν_2). A numerical evaluation of Eqs. (14), (38), and (40) in this limiting case ($\nu_1 = \nu_m$) yields

$$(\bar{\nu}_2 - \bar{\nu}_1)_{\text{max}} - (\bar{\nu}_2 - \bar{\nu}_1)_{\text{thresh}} \cong 1.3[h_0 + 0.55H_0 - 60] \text{ kc/sec} \tag{44}$$

for the parameters in the present maser. The bracketed quantities are again given in Mc/sec. If we assert that the extreme 30 kc/sec shift of the beat reported in Sec. 3 occurs in the second limiting case, Eqs. (43) and (44) imply that $h_0 \cong 40$ Mc/sec and $H_0 \cong 80$

Mc/sec. These values obviously do not represent precise measurements. However, they are sufficiently close to the estimates given in Secs. 10 and 11 to indicate that the hole repulsion effect is the correct explanation for the anomalous power dependence of the beat reported in Sec. 3.

Finally, it is important to note that Eq. (26) also predicts a splitting of the $c/2L$ beat which is compatible with the magnitudes observed. Evaluation of Eq. (26) for three asymmetrically placed holes yields

$$(\bar{\nu}_2 - \bar{\nu}_1) - (\bar{\nu}_1 - \bar{\nu}_3) \cong \frac{3}{8}\frac{\Delta\nu_c}{(c/2L)}\left[\left(\frac{g_2 - f}{f}\right)H_2 - \left(\frac{g_3 - f}{f}\right)H_3\right] - 1.13\,\Delta\nu_c\left(\frac{g_m}{f}\right)\left[\sin\left(\frac{c/2L}{0.6\,\Delta\nu_m}\right)\right]^2 \sin\left(\frac{\nu_m - \nu_1}{0.3\,\Delta\nu_m}\right), \quad (45)$$

where no terms contained in Eqs. (13) and (26) have been neglected. g_2 and H_2 are the fractional energy gain and hole width at ν_2, respectively; g_3 and H_3 are similarly defined at ν_3; and g_m is the fractional energy gain per pass at the line center. Equation (45) holds only for the case where $g_3 \geqq f$ and the gain at the fourth cavity resonance is below threshold for oscillation. Equation (45) predicts that the beat splitting will vanish for $\nu_m = \nu_1$, since in this case $g_2 = g_3$ and $H_2 = H_3$.

In spite of its formidable appearance, Eq. (45) is a fairly linear function of $(\nu_m - \nu_1)$. The main nonlinear term arises from the power dependent part of the hole width in Eq. (38) and enters Eq. (45) as the difference between the squares of two small quantities. For the same reason, Eq. (45) is not strongly dependent on the parameter H_0 at low powers. The splitting is primarily determined by the last term of Eq. (45). Its maximum value occurs at $\nu_m - \nu_1 = \frac{1}{2}(c/2L)$ and at a value for g_m such that the fourth cavity resonance is just below threshold. For the present maser, the maximum value of the second term is a little over 50 kc/sec and hence somewhat in excess of the maximum splitting ($\cong$30 kc/sec) observed. By definition [Eqs. (38) and (39)], the first term in Eq. (45) is positive and, hence subtracts from the non-linear phase characteristic of the Gaussian. At low powers this subtracted term is mainly determined by the width, h_0 in Eq. (38). For the present maser, the inclusion of h_0 in Eq. (45) would reduce the maximum splitting by about 0.5 kc/sec per Mc/sec of h_0. Hence for the value of h_0 estimated above ($\cong$40 Mc/sec), the first term of Eq. (45) would reduce the maximum splitting from the second term to about that value observed. However, the neglect of the possibility of a slight variation in the mirror reflectance coefficient with polarization may render this close agreement fortuitous. The second term in Eq. (38) will certainly increase with the power and at some point may cancel the nonlinearity responsible for the splitting in Eq. (45). It seems evident from the experimental results (Fig. 3) that this limit is not reached in the present case until more than three modes oscillate.

The evaluation of Eq. (45) for more than three holes is straight forward and not terribly interesting. As the nonlinearity contained in Eq. (45) implies, the splitting increases with the number of modes which go into oscillation. In general, the number of components at $n(c/2L)$ is $(N - n)$ where N is the number of holes ($N \geqslant n$) and these components will not be spaced with perfect evenness. In each case, the absolute frequency of the component increases with the pumping rate if the hole repulsion effect predominates. For a symmetric distribution of holes about the Doppler line, the number of components at $n(c/2L)$ is reduced.

14. Conclusion

There exists one perplexing aspect of the hole burning interpretation which should be mentioned. At the highest power levels shown in Fig. 3, the odd-symmetric modes also go into oscillation, as evidenced by the appearance of the 1-Mc/sec satellites in Fig. 2. (These satellites are also split by amounts in the general order of 20 kc/sec.) A question naturally arises regarding the mechanism by which two modes separated by an amount substantially less than the natural linewidth can go into oscillation simultaneously.

We note, however, that this question exists regardless of the interpretation of the beat splitting. It seems probable that the answer to this question lies in the different spatial distributions of the two modes.

The main interest in the beat splitting phenomenon, of course, arises from the possibility of using this phenomenon to make an absolute frequency standard in the optical range. As mentioned in Sec. 3, this might be accomplished by the generation of an error frequency in the case of three oscillating modes and the use of this error frequency to adujst the plate separation. The central resonance may then be filtered out with a suitably designed (low quality) Fabry–Perot interferometer. It seems likely from the data in Sec. 3 that this error frequency would be in the general order of about 0.5 kc/sec per Mc/sec separation of the central cavity resonances from the center of the Doppler line and that the sharpness[2,11] of the maser oscillation would not be a prime limitation on the absolute stability obtainable in practice. Throughout the present analysis it has been assumed that the line shape for the transition is symmetric. This assumption will obviously break down at some point, and it is likely that the ultimate limit would be determined by pressure-dependent effects. The presence of other isotopes of neon can introduce such line asymmetries and it would obviously be desirable to isolate one of these isotopes for use in such a standard. From Thomas' numerical calculations and the natural abundance of Ne^{22} in Ne^{20}, the main effect of this line asymmetry would be to introduce a power-dependent frequency shift at the Doppler center in the order of 30 kc/sec. The presence of other isotopes of neon will also result in slight changes in the nonlinear part of the phase characteristic. In this connection it should be mentioned that no significant difference was found in either the mode pulling or beat splitting effects when the present maser was operated on a neon sample containing 99.9% Ne^{20} as opposed to neon samples containing the normal isotopic abundance.

There are two more important sources of practical difficulty in utilizing the beat splitting effect to determine the absolute frequency of the central mode in respect to the Doppler center. The first consists of the possibility that the mirror reflectance coefficient may vary slightly with polarization.[8] If two modes were polarized in one direction and the third orthogonally, this effect could introduce an additional shift in the beat splitting. The effect enters by changing the hole depth and therefore the magnitude of the hole repulsion term. From Eq. (45) the shift in the splitting would amount to about $\frac{3}{8}\Delta\nu_c(2LH/c)\Delta f/f$, where Δf is the extreme variation of the fractional loss between the two orientations and H is the width of the hole for the orthogonally polarized mode. For example, for a hole width of 100 Mc/sec, a 1% variation in the mirror losses between the two extreme orientations would shift the beat splitting by about 2 kc/sec. Second, it has been found by others[9,10] using different observational techniques that the difference frequency between the two $c/2L$ components [Fig. 3(b)] jumps discontinuously to zero at some point below about 1 kc/sec. Under these conditions the $2c/2L$ and $c/2L$ beats are both harmonically related[9] and phase-locked[10] to a very high degree. Hence one completely loses track of the absolute frequencies at this point and it seems apparent that a minimum absolute frequency uncertainty in the order of 2 Mc/sec will exist. This effect is definitely not contained in Eq. (45). Although Eq. (45) does contain a slight nonlinear dependence on $(\nu_m - \nu_1)$, the nonlinear terms are much too small to give, for example, a cubic equation with real roots or a point of inflection near the origin. The effect probably has its origin in nonlinear time-dependent properties of the medium, which, as previously stated, have been neglected for reasons of simplicity.

Neither of these effects would prevent one from using the beat splitting phenomenon to obtain a high degree of relative frequency stabilization when the central mode is slightly detuned from the Doppler center.

Acknowledgments

The author is particularly indebted to A. J. Rack for the use of his frequency measuring apparatus and help in obtaining the data in Table 1. It is also a pleasure to acknowledge the help of D. E. Thomas for his numerical phase shift calculations and the technical assistance of P. Kindlmann. The author is also indebted to Professor W. W. Watson of Yale University for supplying the highly pure Ne^{20} sample used. In addition, the author has benefitted from helpful discussions with Dr. W. L. Faust, Dr. C. G. B. Garrett, Dr. A. Javan and Dr. J. A. White.

Appendix I

The phase shifts associated with a given line shape may be calculated most generally using methods equivalent to the Kramers–Kronig relations.[15] The assumptions on the following derivation are that the process obeys causality (i.e., the amplitude gain function is analytic in the lower half of the complex ω plane), that the gain is small, and that the running wave is amplified linearly as it traverses the cavity. Although the gain saturates in the system during oscillation, we assume that amplification by this saturated gain characteristic is still linear.

We let $K(\omega)$ be the complex, fractional amplitude gain in the medium defined by

$$1 + K(\omega) = [1 + K_0(\omega)] \exp[-i\phi(\omega)], \tag{I.1}$$

where $K_0(\omega)$ is the fractional amplitude gain per pass and $\phi(\omega)$ is the additional single-pass phase shift resulting from the amplifying medium. Taking the logarithm of (I.1) and expanding in the limits $K(\omega), K_0(\omega) \ll 1$,

$$K(\omega) \cong K_0(\omega) - i\phi(\omega). \tag{I.2}$$

We note at this point that the fractional energy gain per pass, $g_0(\omega)$, is

$$g_0(\omega) \cong 2K_0(\omega). \tag{I.3}$$

We assume $K(\omega)$ vanishes at $\omega = \pm\infty, \pm i\infty$, and has no singularities in the lower half-plane. Then

$$\oint \frac{K(\omega)\, d\omega}{\omega - \omega_0} = 0, \tag{I.4}$$

where the integral is taken along the real axis, skirts under the pole at ω_0 and is closed in the lower half of the complex ω plane. Substituting (I.2) and equating real parts,

$$\phi(\omega_0) = -(1/\pi)\int_{-\infty}^{+\infty} \frac{K_0(\omega)\,d\omega}{\omega - \omega_0}, \tag{I.5}$$

where (1.5) represents the principal part of the integral on the real axis. Equation (I.5) is more convenient for the Lorentzian than the usual form of the Kramers–Kronig relations deduced from it.

We take for the Lorentzian,

$$K_0(\omega) \cong K_m(\Delta\omega)^2[(\Delta\omega)^2 + 4\,(\omega_m - \omega)^2]^{-1} \tag{I.6}$$

where the approximation $\omega + \omega_m \cong 2\omega$ has been made. Although $K_0(\omega)$ has a pole in the lower half-plane, $K(\omega)$ does not. We next take,

$$I = \oint \frac{K_0(\omega)\,d\omega}{\omega - \omega_0}, \tag{I.7}$$

where the path of integration runs along the real axis, skirts under the pole at ω_0 and is closed in the upper half-plane. From (I.5), (I.6), and (I.7),

$$-\pi\phi(\omega_0) + \pi i K_0(\omega_0) = 2\pi i \sum(\text{Residues}). \tag{I.8}$$

Hence,

$$\begin{aligned}\phi(\omega_0) &= -2K_0(\omega_0)(\omega_m - \omega_0)/\Delta\omega \\ &= -g_0(\omega_0)(\omega_m - \omega_0)/\Delta\omega \end{aligned} \tag{I.9}$$

yielding Eq. (11) of the text.

Appendix II

The hole widths may be evaluated through a standard application of time-dependent perturbation theory.[20] We assume that

[20] A similar analysis for levels having different decay rates has been given for the ground state of positronium by V. W. Hughes, S. Marder, and C. S. Wu [*Phys. Rev.* **106,** 934 (1957), Appendix I]. The approximation, $\gamma_b \gg \gamma_a$, appropriate to the latter reference has, however, not been made here.

the two maser levels are described by the wave function

$$\Psi(r, t) = A\psi_a + B\psi_b. \tag{II.1}$$

Satisfying

$$(H_0 + H')\Psi = i\hbar\, \partial\Psi/\partial t \tag{II.2}$$

in the presence of the electromagnetic field. H_0 is the original Hamiltonian and H' the time-dependent perturbation. The subscript a denotes the upper maser level and b the lower. The probability amplitudes A and B are functions of the time. We take the electric field to be linearly polarized in the z direction. Hence,

$$H' = -ezE_0 \cos \omega t = -(ezE_0/2)(e^{i\omega t} + e^{-i\omega t}). \tag{II.3}$$

Since the perturbation is electric dipole in nature, only off-diagonal matrix elements arise. Neglecting the anti-resonant term and the interaction with all neighboring levels,

$$\begin{aligned} \dot{A} &= (V/i\hbar)e^{-i(\omega-\omega_0)t}B - (\gamma_a/2)A, \\ \dot{B} &= (V/i\hbar)e^{i(\omega-\omega_0)t}A - (\gamma_b/2)B, \end{aligned} \tag{II.4}$$

where the spontaneous decay rates γ_a and γ_b of the upper and lower maser levels have been introduced phenomenologically. From (II.3)

$$V = -\tfrac{1}{2}eE_0(z)_{a,b}. \tag{II.5}$$

ω_0 represents the resonant frequency of the atomic transition and ω the frequency of the oscillating field.

We seek those solutions to the coupled Eqs. (II.4) satisfying the initial conditions

$$A = 1 \quad \text{and} \quad B = 0 \quad \text{at} \quad t = 0. \tag{II.6}$$

Condition (II.6) corresponds to putting the atom in the upper state at $t = 0$. The total probability that the atom decays by stimulated emission is then given by

$$P_s = \gamma_b \int_0^\infty |B(t)|^2\, dt. \tag{II.7}$$

Through standard methods, the solution of Eqs. (II.4) for $B(t)$ subject to initial conditions (II.6) is

$$B(t) = \frac{i\hbar}{V} \frac{(\mu_- + \frac{1}{2}\gamma_a)(\mu_+ + \frac{1}{2}\gamma_a)}{(\mu_+ - \mu_-)} [e^{\mu_- t} - e^{\mu_+ t}] e^{i(\omega - \omega_0)t}, \qquad \text{(II.8)}$$

where

$$\begin{aligned} \mu_\pm = & -\tfrac{1}{2}[\tfrac{1}{2}(\gamma_a + \gamma_b) + i(\omega - \omega_0)] \\ & \pm \tfrac{1}{2}\{[\tfrac{1}{2}(\gamma_b - \gamma_a) + i(\omega - \omega_0)]^2 - |2V/\hbar|^2\}^{\frac{1}{2}}. \end{aligned} \qquad \text{(II.9)}$$

Hence

$$P_s - \gamma_b \left|\frac{\hbar}{V}\right|^2 \left|\frac{(\mu_- + \frac{1}{2}\gamma_a)(\mu_+ + \frac{1}{2}\gamma_a)}{(\mu_+ - \mu_-)}\right|^2 \times \int_0^\infty |e^{\mu_- t} - e^{\mu_+ t}|^2 \, dt. \qquad \text{(II.10)}$$

After some algebraic manipulation, Eq. (II.10) reduces to

$$P_s = \frac{\gamma_b(\gamma_a + \gamma_b)|2V/\hbar|^2}{4(\omega - \omega_0)^2 \gamma_a \gamma_b + (\gamma_a \gamma_b + |2V/\hbar|^2)(\gamma_a + \gamma_b)^2}, \qquad \text{(II.11)}$$

yielding Eq. (27) of the text.[21] Aside from the minor approximations made in formulating Eqs. (II.4) this expression for the probability function is exact.

[21] *Note added in proof.* A result equivalent to Eq. (II.11) has been obtained from the density matrix method by W. E. Lamb, Jr., and T. M. Sanders, Jr., *Phys. Rev.* **119**, 1901 (1960).

Index